Transactions on Computer Systems and Networks

Transactions on Computer Systems and Networks is a unique series that aims to capture advances in evolution of computer hardware and software systems and progress in computer networks. Computing Systems in present world span from miniature IoT nodes and embedded computing systems to large-scale cloud infrastructures, which necessitates developing systems architecture, storage infrastructure and process management to work at various scales. Present day networking technologies provide pervasive global coverage on a scale and enable multitude of transformative technologies. The new landscape of computing comprises of self-aware autonomous systems, which are built upon a software-hardware collaborative framework. These systems are designed to execute critical and non-critical tasks involving a variety of processing resources like multi-core CPUs, reconfigurable hardware, GPUs and TPUs which are managed through virtualisation, real-time process management and fault-tolerance. While AI, Machine Learning and Deep Learning tasks are predominantly increasing in the application space the computing system research aim towards efficient means of data processing, memory management, real-time task scheduling, scalable, secured and energy aware computing. The paradigm of computer networks also extends it support to this evolving application scenario through various advanced protocols, architectures and services. This series aims to present leading works on advances in theory, design, behaviour and applications in computing systems and networks. The Series accepts research monographs, introductory and advanced textbooks, professional books, reference works, and select conference proceedings.

Pushpendu Kar · Monideepa Roy · Sujoy Datta

Recommender Systems: Algorithms and their Applications

Springer

Pushpendu Kar
School of Computer Science
University of Nottingham Ningbo China
Ningbo, China

Monideepa Roy
School of Computer Engineering
KIIT Deemed University
Bhubaneswar, Odisha, India

Sujoy Datta
School of Computer Engineering
KIIT Deemed University
Bhubaneswar, Odisha, India

ISSN 2730-7484 ISSN 2730-7492 (electronic)
Transactions on Computer Systems and Networks
ISBN 978-981-97-0540-5 ISBN 978-981-97-0538-2 (eBook)
https://doi.org/10.1007/978-981-97-0538-2

This Springer imprint is published by the registered company Springer Nature Singapore Pte Ltd.
The registered company address is: 152 Beach Road, #21-01/04 Gateway East, Singapore 189721, Singapore

If disposing of this product, please recycle the paper.

Preface

Recommendation systems were introduced in the 90's but have gradually become an indispensable tool with the advent of numerous e-commerce companies. Recent years have seen a huge jump in the number of such web services, and they rely heavily on recommendation systems to gain an advantage over their competitors. Recommendation systems gather information about the likes and dislikes of a user and use various types of complex algorithms to predict what a user may be interested in and send personalized recommendations to users. Brands like Netflix, Amazon, Facebook, Spotify, and YouTube collect information about users and try to predict user preferences. If a person buys a certain product, then suggestions for similar products are sent to the user. If a user likes a particular type of music or movie, then it will try to predict and recommend similar types of music or movies to the user. It is a very vast and interesting area of research but at present, in this book, we have taken some of the most important topics which form the basis of recommender systems, along with some case studies and applications and suggestions for future research directions.

This book will be useful to users who are new to the topic and wish to learn it. It will also be useful to advanced users who know the theory but want to implement or design a system from scratch and can learn from the different types of algorithms.

This book consists of 12 chapters.

Chapter 1 is a general introduction of what is the importance of recommender systems and an overview of the scope of the book and its audience and the motivation behind writing this book.

Chapter 2 is a general overview of all possible types of algorithms for recommendation systems.

Chapter 3 discusses two of the most widely used types of recommender algorithms, content-based systems and collaborative filtering methods, and their features and suitability for implementation.

Chapter 4 discusses the decomposition of the matrix in clustering.

Chapter 5 discusses how to learn to rank users based on various factors and how to detect profiles of false users, along with the Shilling attack example.

Chapter 6 deals with knowledge-based, ensemble-based, and hybrid recommender systems.

Chapter 7 discusses how to deal with the big data associated with recommender systems.

Chapter 8 discusses the existing trust-centric and attack-resistance techniques for recommender systems and proposes different ways to improve the performance of recommendation systems based on both attack and trust.

Chapter 9 shows the steps in building a recommendation engine.

Chapter 10 discusses different types of healthcare recommendation systems, challenges, and the scope of improvements.

Chapter 11 discusses the application of recommender systems to military surveillance.

Chapter 12 discusses the use of recommender systems in different real application domains, existing challenges as well as the scopes and ideas of their improvements.

Ningbo, China Pushpendu Kar
Bhubaneswar, India Monideepa Roy
Bhubaneswar, India Sujoy Datta

Acknowledgments

I would like to thank my parents, **Mihir Kumar Kar** and **Pratima Kar**, my wife, **Sangita**, my daughter, **Ritosmita**, and my son, **Ritanshu**, for their continuous support, guidance, and encouragement. I would like to express my sincere gratitude to **Tianyi Ma** and **Chenyu Yang** for helping in writing Chap. 8, **Zhihang Zhu** for helping in writing Chap. 10, and **Xinyi Wang** for helping in writing Chap. 12.

—Pushpendu Kar

I would like to express my sincere thanks to my Mom **(late Hasi Roy)** and Dad **(late Sunil K. Roy)** for their blessings even when they are no more physically here to guide me, but I'm sure they are watching with satisfaction from above. I would like to thank my sister **Madhumita Roy** for her constant support and inspiration throughout this assignment. I am also thankful to her for kindly designing the cover for the book. I would like to thank my B.Tech. students **Aishi Paul** and **Divyansi Mishra** for their help in drawing the diagrams for the book. Thanks to each and every one of you for your timely support and help.

—Monideepa Roy

I would like to express my sincere thanks to my father **(Sushil Dutta)** and my mother **(late Mintu Dutta)** for their constant inspiration and blessings. I would also like to thank my sisters **Sutapa Rakshit** and **Sunanda Rakshit** for their continuous encouragement and support to me during this assignment.

—Sujoy Datta

Contents

About the Authors

Dr. Pushpendu Kar is currently working as an Assistant Professor in the School of Computer Science, University of Nottingham (China campus). Before this, he was a Postdoctoral Research Fellow at the Norwegian University of Science and Technology, the National University of Singapore, and Nanyang Technological University. He also worked in different engineering colleges as a lecturer and in the IT industry as a software professional. He has more than 12 years of teaching and research experience as well as one and a half years of industrial experience at IBM. He has completed all his Ph.D., Master of Engineering, and Bachelor of Technology in Computer Science and Engineering. He also completed Postgraduate Certificate in Higher Education (PGCHE) in 2023. He was awarded the prestigious Erasmus Mundus Postdoctoral Fellowship from the European Commission, ERCIM Alain Bensoussan Fellowship from the European Union, and SERB OPD Fellowship from the Department of Science and Technology, Government of India. He has received the 2020 IEEE Systems Journal Best Paper Award. He has received four research grants for conducting research-based projects, three of them as a Principal Investigator (PI). He also received many travel grants to attend conferences and doctoral colloquiums. He is the author of more than 60 scholarly research papers, which have been published in reputed journals, conferences, book chapters, and IT magazines. He has also published three books. He is an inventor of five patents. He has chaired several conference committees, worked as a team member to organize short-term courses, and delivered a few invited talks as well as Keynote Lectures at international conferences and institutions. He is a Senior Member of IEEE, a Senior Fellow of the Higher Education Academy (SFHEA), UK, and Fellow of the Institution of Electronics and Telecommunication Engineers (FIETE), India. Ningbo Municipal Government, China has recognized him as a High-Level Talent. His research areas include Wireless Sensor Networks, Internet of Things, Content-Centric Networking, Machine Learning, and Blockchain.

Dr. Monideepa Roy did her Bachelors and Masters in Mathematics from IIT Kharagpur, and her Ph.D. in CSE from Jadavpur University. Currently she is working as an Associate Professor at KIIT Deemed University, Bhubaneswar since the last 11 years. Her areas of interest include Remote Healthcare, Mobile Computing, Cognitive WSNs, Remote Sensing, Recommender Systems, Sparse Approximations, and Artificial Neural Networks. At present she has seven research scholars working with her in the above areas and two more have successfully defended their theses under her guidance. She has several publications in reputed conferences and journals. She has been the Organizing Chair of the first two editions of the International Conference on Computational Intelligence and Networks CINE 2015 and 2016, ICMC 2019 and has organised several workshops and seminars. She also has several book chapter publications in Springer as well as two edited books under Taylor and Francis.

Sujoy Datta has done his M.Tech. from IIT Kharagpur. Currently he is working as an Assistant Professor in the School of Computer Engineering, KIIT Deemed University since the last eleven years. His areas of research include Wireless networks, Computer Security, Elliptic curve cryptography and neural networks, Remote Healthcare and Recommender Systems. He has several publications in various conferences and journals. He has co-organised several workshops and international conferences as well as several workshops and seminars. He also has several book chapter publications in Springer as well as two edited books by Taylor and Francis.

Chapter 1
Introduction to Recommendation Systems

Abstract With the rapid growth of e-commerce, the web has become a very popular source of doing business by various companies. Customers also find it a very attractive proposition as it saves the time to go outside and shop for what a user needs, as well as the fact that users have access to a huge array of choices to buy from. Since it is a very tough and competitive market, and companies have realized that people usually tend to buy similar types of products or watch similar types of movies, they have now resorted to modern technology to make it easier for customers to make their choices. This led to the advent of various recommendation algorithms with which the companies are now able to predict the choices and personal preferences of their customers and accordingly push appropriate suggestions or recommendations for products that a person is likely to purchase. With the huge success of recommendation systems, they are now widely being adopted by more and more brands and for more varieties of applications. This chapter gives an overview of the reasons why recommendation systems have become so popular.

Keywords Recommender algorithms · Product recommendations · Machine learning · Customer choices · Prediction

1.1 Introduction

Since consumers today are faced with huge numbers of choices in terms of new products or new movies to watch, and less time on their hands, so it's difficult to make the choices of selection of the most relevant options on their own. So, whenever a person buys a new product or wants to watch a new movie, he/she prefers to find out ratings or recommendations from past users to make their choices faster and easier (Abowd et al. 1999). However, even that is time-consuming with the huge volumes of data. So, this has led to the emergence of recommendation systems, which use algorithms to predict and find the best matches for a person based on various parameters. They also form the basis of many machine learning algorithms. In this book, we take a look at the different types of algorithms that are used for generating

P. Kar et al., *Recommender Systems: Algorithms and their Applications*, Transactions on Computer Systems and Networks, https://doi.org/10.1007/978-981-97-0538-2_1

accurate predictions for consumers and some applications of recommender systems. The rest of the chapter is organized as follows: Sect. 1.2 defines what recommendation systems are, Sect. 1.3 discusses who can benefit from recommendation systems, Sect. 1.4 discusses how that can help, Sect. 1.5 describes how they work, Sect. 1.6 describes the evolution of recommender systems, Sect. 1.7 takes the examples of some famous brands who have used recommendation systems very successfully, Sect. 1.8 describes the scope of the book, Sect. 1.9 gives an overview of the chapters, and Sect. 1.10 is the summary.

1.2 What Are Recommendation Systems?

As defined in Wikipedia, "A recommender system, or a recommendation system (sometimes replacing 'system' with a synonym such as a platform or an engine), is a subclass of information filtering system that seeks to predict the "rating" or "preference" a user would give to an item."

So, the basic aim of a recommender system is to provide users with the most relevant suggestions for things to buy, places to visit, or movies to watch, based on what the user had chosen earlier or what people with similar profiles have chosen. Recommender systems (Adamopoulos et al. 2014) predict the choices of the users and then suggest the most relevant options. In this present age of competition, there is a huge number of choices available to consumers, especially in the e-commerce domain. So to gain an edge over other competitors, a retailer or business needs to be able to correctly predict user preferences and send appropriate suggestions for their products to the consumers. This makes recommendation systems the most powerful machine learning technique which is widely used by online retailers to gain an edge over the others and increase their profits. So how do they actually work and how do they predict the preferences of their users? The data that is required for the recommendation systems to make such predictions is obtained from various sources. Data is collected explicitly from user ratings that are collected after a person has watched a movie or listened to a song or purchased an item, implicitly through search engine queries and purchase histories through cookies, or from past knowledge about the user or the item. Many sites like Netflix, Facebook, Amazon, Spotify, and YouTube use such types of data and implement their algorithms to provide the most relevant suggestions to the users.

1.3 Who Can Benefit from Them?

Although any business can benefit from implementing recommendation systems, the two main factors which determine the extent to which a business can benefit from recommendation systems are:

Breadth of data—If the business has only a few customers, and they behave in different ways, then using an automated recommendation system will not be of much use to them. It will be much easier to let the employees use their own logic to predict the preferences of the individual customers.

Depth of data—When the business has only one single data point for each of their customers, recommendation systems will not have sufficient training data to base their predictions on.

So organizations who can benefit from automated recommendation systems can vary from e-commerce, retail and media to banking, telecom and other utilities. There are of course many more areas which can benefit from implementing recommendation systems, but here we describe some of the most popular ones. A more detailed discussion of specific brands will be done in the later chapters.

E-commerce is one of the first areas where recommendation systems were used. Because these companies have access to online data of millions of customers, they can easily use that data to generate accurate recommendations.

Retail is another area which can benefit to a great extent from recommendation systems. Since retailers have direct access to huge volumes of shopping data therefore they have a very good idea of the customers intent and can make accurate predictions.

Media industry is also one of the first few companies which were the first to implement recommendation systems. Almost all news channels use recommendation engines.

Banking is also a very important application where the financial situations and past preferences of millions of customer data make a very comprehensive data bank.

The telecom industry also has a similar dynamics as that of the banking industry where the service providers have access to a wide variety of customer data, call and usage preferences, and past data of a huge volume of customers. The telecom industry has the additional advantage that it has a limited number of products for which they need information, to make their predictions on (Fig. 1.1).

Fig. 1.1 Increasing importance of personalization in the post pandemic market. Source - McKinsey

1.4 How Do They Help?

The performance of some companies who have implemented recommender systems:

- Cross-selling and category-penetration techniques increase sales by 20% and profits by 30%, according to McKinsey (https://www.mckinsey.com/capabilities/growth-marketing-and-sales/how-we-help-clients/clm-online-retailer).
- 35% of the purchases on Amazon are the result of their recommender system, according to McKinsey (https://exposebox.com/the-power-of-product-recommendation2022/#:~:text=According%20to%20an%20article%20by,third%20of%20its%20product%20sales) (March 6, 2022).
- During the Chinese global shopping festival of November 11, 2016, Alibaba achieved growth of up to 20% of their conversion rate using personalized landing pages, according to Alizila (https://www.alizila.com/live-updates-alibabas-11-11-global-shopping-festival/) (November 10, 2016).
- Recommendations are responsible for 70% of the time people spend watching videos on YouTube (https://blog.hootsuite.com/how-the-youtube-algorithm-works/) (April 18, 2023).
- 75% of what people are watching on Netflix comes from recommendations, according to McKinsey (https://www.mckinsey.com/industries/retail/our-insights/how-retailers-can-keep-up-with-consumers).
- Employing a recommender system enables Netflix to save around $1 billion each year, according to an article (https://www.businessinsider.in/tech/why-netflix-thinks-its-personalized-recommendation-engine-is-worth-1-billion-per-year/articleshow/52754724.cms).

All major companies are now aware that they need to reach their customers and advertise their products before others to create an impression on the customers and influence them to buy their products. So they are increasingly depending on recommendation systems to increase their sales and profits by providing the best possible personalized product suggestions and a better consumer experience based on consumer preferences. Recommendations also help the consumers by speeding up the searches and making it much easier for them to get access content of their interest, and providing offers that the users might not have normally searched for or known. In this way, the companies are also gaining and retaining customers through sending emails containing links to the latest offers that are of interest to the users or suggesting movies or web series, or restaurants that the user is likely to watch or visit respectively.

This in turn creates a sense of dependency and loyalty among the customers and they start feeling that their preferences are retained and known to the system, and they will be more likely to use that platform in the future for buying more products or viewing more content if they had an enhanced shopping experience earlier. So, the chances of the customer going to another platform actually decreases because by nature people are slow to change if they are already comfortable with a particular system. So, the threat of losing a customer to others also decreases.

So accurate recommendations add value to the credibility of a company and make their products more appealing to the users if they have more online presence in searches and companies can leverage it to capture more customers and increase their sales (Adomavicius and Tuzhilin 2011).

So the broad benefits of a recommendation system can be summarized as follows:

- *Increased sales/conversion*—Usually, to increase sales, one also needs to increase their marketing efforts. But if a company already has a good automated recommendation system in place, they can achieve additional recurring sales without extra effort.
- *Increased user satisfaction*—It leads to increased customer satisfaction, because it reduces the customers path time to sales, by making appropriate recommendations to them even before they search for it.
- *Increased loyalty and share of mind*—When customers spend more time on the website of a particular brand, they get more familiar with it therefore increases the probability of making future purchases with that brand.
- *Reduced churn*—Emails which are powered by recommendation systems are a very good way to re-engage customers and reduce churn or attrition from one brand to another. Discounts or coupons are yet another effective way to bring in customers and coupled with accurate recommendations can increase the probability of the conversion of a customer.

1.5 How Do They Work?

Recommender systems need two types of information that they can work on. One is characteristics information which is information about items (like keywords and categories) and users (like preferences, profiles). The system will find out the personal preferences of users and maintain a profile of their customers so that they can make suggestions to new users with similar profiles. The other is user–item interactions like ratings, number of purchases, likes, etc. where the user rates a product he/she has experienced. Based on this the algorithms in recommender systems can be broadly divided into three categories: content-based, collaborative filtering, and hybrid systems (Adomavicius et al. 2011). The content cased systems use characteristics information while the collaborative filtering uses interactions between the users and the items. Hybrid systems are a combination of both of them. A more detailed study of the above types of algorithms is given in Chap. 3.

1.6 The Evolution of Recommender Systems

The first recommendation engine was made in 1992 at the Xerox Palo Alto Research Center. It was mainly designed with the purpose of allowing users rate the messages /documents of an experimental mail system called Tapestry. It used a method called collaborative filtering to for the recommendation engine to tell the user about which were the most read or most loved documents. It proved to be an efficient process and gave very good results for tapestry. It was later developed further to be able to perform more complex operations like filtering, retrieval, and browsing of e-documents. Figure 1.2 shows the three generations of recommendation engines.

The most successful recommendation engine was probably built by Amazon, when it made it to the list of the top 10 retailers in 2012. Amazon was at 10th position with a revenue of 34.4 billion USD. As per McKinsey, 35% of Amazon's revenue in 2012 came from recommendation engines.

The reason their recommendation engine was so successful as compared to the others was that it adapted to the challenges that came with an increase in the number of customers. So instead of building focus on each customer individually and giving them recommendations based on their past activities, what they did was to make clusters of customers who had similar choices. As a result they found out that the end results were more accurate and that the email recommendations were the best way to convince a customer to buy their product.

Content-based filtering is a method which focuses on the likes and dislikes of a single user and preferences. The collaborative filtering method focuses more on analyzing the preferences of a group of people.

A hybrid filtering method utilizes both the above two methods and focuses more on what a customer might need instead of what he/she wants.

So how does one choose the most suitable recommendation engine? (Adomavicius and Kwon 2007).

There are primarily two factors to be considered:

The first point is why does a particular business need a recommendation engine. If it has a loyal customer group with a limited number of people, then content filtering is good enough. However if there is some complexity in the data then either collaborative filtering or hybrid filtering should be chosen.

The second point is whether there will be a need to scale up in the future. If there is a possibility of scale up, then the recommendation engine should be chosen accordingly.

The Three Generation of Recommendation Engine

First Tier	Second Tier	Third Tier
Mainly used in the e-commerce and SaaS industry. It Uses content-based, collaborative- based and hybrid filtering algorithms. Recommendations are more generic.	The second Tier is an advancement of the first tier engine that uses social information to create user portfolios, which helps make the recommendations more personalized.	The third-tier recommendation system is the smartest as it uses artificial intelligence to know the location and emotional factors. This knowledge-based recommender system gives by far the most precise recommendations to users.

Fig. 1.2 Three generations of recommendation engines

1.7 Some Major Brands Who Are Very Successfully Based on Recommendation Systems

Amazon.com—Amazon uses a method called item to item collaborative filtering for generating recommendations on most of their pages of their website. According to McKinsey, a major 35% of its purchases are as a result of recommendations. In Fig. 1.3 we see a sample page which has customized recommendations for a particular user.

Netflix—Netflix is another company which is very highly data driven, which utilizes recommendation systems to boost customer satisfaction. Again as per a study performed by McKinsey, a whopping 75% of Netflix viewing is driven by recommendations. A more detailed discussion of the Netflix case study will be done in the later chapters. Figure 1.4 shows a sample screenshot of the movie or web series recommendations that are sent to a viewer based on the past viewing history.

Spotify—Spotify software engineer Edward Newett has developed a new system to help Spotify users to discover new music and the tool is called Discover Weekly, Fig. 1.5. It was launched around 7 years ago and now has over 40 million users. Every week Spotify generates customized new playlists for each of its customers, which is a list of 30 songs based on the unique preferences of the customer. They acquired Echo Nest, which was a music intelligence and data analytics startup and created a music recommendation engine which uses three types of recommendation models—collaborative filtering, natural language processing, and audio file analysis.

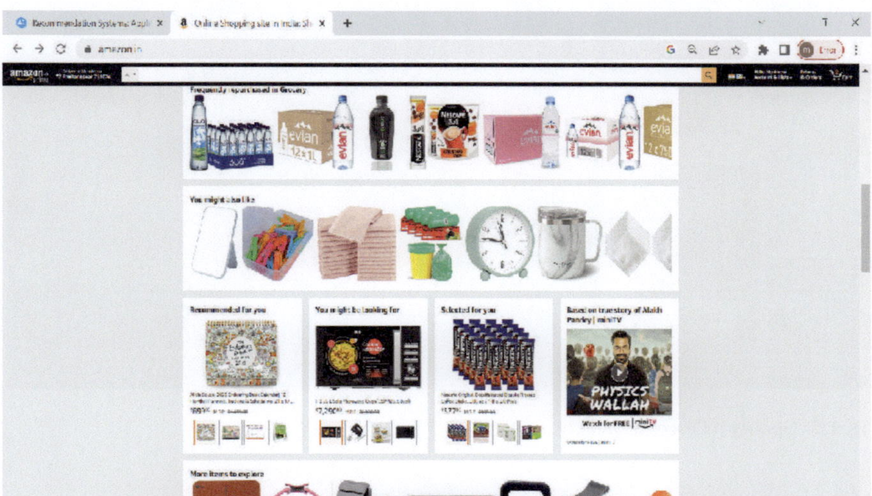

Fig. 1.3 A sample Amazon shopping page

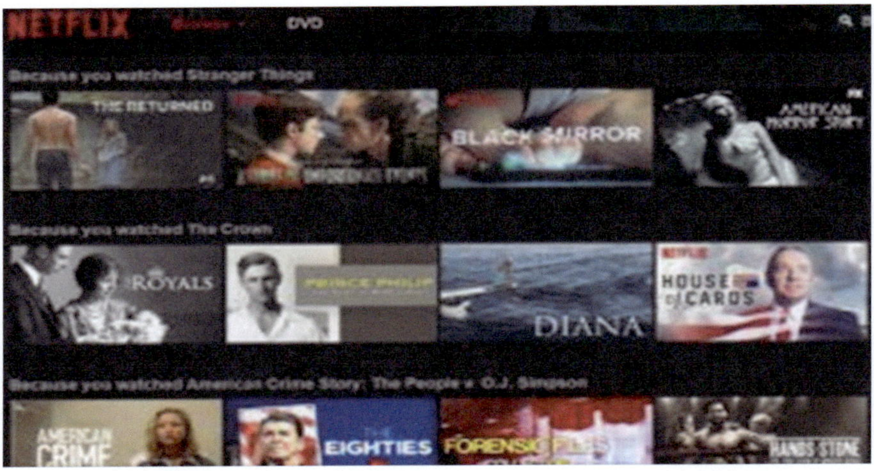

Fig. 1.4 A sample Netflix recommendation page

Fig. 1.5 Spotify's Discover weekly

LinkedIn—Like many other social media channels, LinkedIn also uses "you may also know" types of recommendations, and the number of common connections between any two persons.

1.8 Scope of the Book

This book is mainly aimed as a hands-on help to those who are learning for the first time and wish to learn about recommendation systems from scratch and want to implement them for some applications. The book gives an overview of all the different types of algorithms and application scenarios. It also deals with security and the types of attacks that are usually faced by a recommendation system and ways to detect and safeguard a system from it. There are also two case studies to show the importance and widespread applicability of recommendation systems. There is also a chapter that discusses some novel and diverse applications of recommender systems.

1.9 Overview of the Chapters

The book consists of eleven other chapters. Chapter 2 gives a broad overview of all types of recommendation algorithms. Chapter 3 deals with the three main types of algorithms, collaborative filtering, clustering, and hybrid algorithms. Chapter 4 is about the decomposition of matrices for clustering. Chapter 5 deals with the problem of how to rank the choices correctly and how to safeguard against attacks and fake profiles. Chapter 6 is about knowledge-based and ensemble-based recommender systems. Chapter 7 deals with the big data behind recommender systems. Chapter 8 discusses the importance of trust-centric and attack-resistance recommender systems. Chapter 9 shows the steps in building a recommendation system. Chapters 10 and 11 are applications of recommender systems in healthcare and surveillance, respectively. Chapter 12 discusses some novel applications of recommender systems as well as scopes and ideas for their improvement.

1.10 Summary

In this chapter, we have seen why recommender systems have become so popular and why a study of the various types of algorithms is so important to get a clear understanding of how the process of recommendation works. It defines the scope of this book and gives an overview of its contents of the book. The next chapter is a formal introduction to recommendation systems and gives an overview of the various types of algorithms used.

Think Tank

1. *What are the broad benefits of using a recommender system?*
2. *What factors should be kept in mind while selecting the most suitable recommendation algorithm for a particular job?*
3. *Which types of business can benefit from using recommendation systems?*
4. *In which circumstances will recommendation systems NOT provide accurate predictions?*

References

Abowd G, Dey A, Brown P, Davies N, Smith M, Steggles P (1999) Towards a better understanding of context and context-awareness. In: Gellersen H-W (ed) Handheld and ubiquitous computing. Springer, Berlin, pp 304–307

Adamopoulos P, Bellogin A, Castells P, Cremonesi P, Steck H (2014) REDD 2014—International Workshop on Recommender Systems Evaluation: Dimensions and Design. Held in conjunction with ACM Conference on Recommender systems

Adomavicius G, Tuzhilin A (2011) Context-aware recommender systems. In: Ricci F, Rokach L, Shapira B, Kantor PB (eds) Recommender systems handbook. Springer, New York, pp 217–253

Adomavicius G, Manouselis N, Kwon Y (2011) Multi-criteria recommender systems. In: Ricci F, Rokach L, Shapira B, Kantor PB (eds) Recommender systems handbook. Springer, New York, pp 769–803

Adomavicius G, Kwon Y (2007) New recommendation techniques for multi-criteria rating systems. IEEE Intell Syst 22(3):48–55

Chapter 2
Overview of Recommendation Systems

Abstract In this chapter, an overview of the recommendation algorithms is given. The chapter deals with the goals of a recommender system and discusses the wide spectrum of applications they can be applied for. Some very successful business models which run on recommendation systems have also been discussed here. After that a classification of the different types of recommendation systems is given, followed by the specific challenges faced for domains.

Keywords Content based · Collaborative filtering · Personalized systems · Hybrid · Cold start · Explicit feedback

2.1 Introduction

With more and more companies using the Web as a medium for their business, recommendation systems have become a very important tool for them to keep themselves ahead of their competitors by providing the best-personalized recommendations for a wide variety of items to users (Adomavicius et al. 2005; Adomavicius and Tuzhilin 2005a). One of the main factors why recommendation systems have become so popular is that it is very easy to get user feedback about an item through online services. A company can easily find out the likes and dislikes of a person through various feedback mechanisms. For example on Netflix, a user can easily rate a movie that he/she has watched just by simply clicking on the mouse. These are explicit feedbacks where a user can give ratings in numerical or other formats. But there are implicit types of feedback systems also. Even when a user is simply browsing for some products, it usually means that the customer is interested in such types of products. These types of feedback are used by online sellers like Amazon, Nykaa, TataCliq, etc., and the data is also collected very effortlessly in terms of activity done by the customer. So recommender systems basically utilize such types of data to predict customer preferences.

P. Kar et al., *Recommender Systems: Algorithms and their Applications*, Transactions on Computer Systems and Networks, https://doi.org/10.1007/978-981-97-0538-2_2

2.2 Goals of Recommender Systems

The most common goals of a recommender system (Adomavicius and Tuzhilin 2005b; Averjanova et al. 2008) are as follows:

- *Relevance*—The first goal of a recommender system is to be able to make accurate predictions about the personal preferences of customers and suggest the relevant products of their interest to them. The main logic behind this is a person has a higher probability of consuming an item that is of his/her choice or preference. However, there are other secondary goals also.
- *Novelty*—Recommender systems are really useful if they are successful in suggesting products of interest to the user when the user has not seen the product in the past. So if a user likes a specific genre of movies, then the recommender system can make suggestions for the new arrivals.
- *Serendipity*—Since sometimes some of the products suggested by a system are somewhat unexpected to a customer, therefore there is an element of a lucky discovery here also and may actually surprise the user, rather than just being something that the user does not know about.
- Increasing recommendation diversity - usually, recommender systems suggest the list of the top k items in a list. But if all systems return the same lists then it becomes more probable that the user might not like any of them. On the other hand, if there are diverse suggestions, then there is a higher chance that a user will be liking at least one of these items.

2.3 The Spectrum of Applications

There are a variety of applications that use recommender systems at present. Some famous brands are discussed (Baccigalupo and Plaza 2006; Bailey 2008).

Amazon.com recommender system

Amazon was one of the pioneers in recommender systems and realized its benefits very early. It's an online retailer brand and sells a variety of products through its web portal. It started originally as a book retailer but slowly progressed to many other categories like software, electronics, games, tools, gifting, household, movies, cosmetics, food, etc. Amazon has an explicit rating facility where it allows the user to rate items on a 5-point scale. It also tracks the buying behaviors, previous items purchased, and browsing history.

Netflix Movie Recommender System

Netflix is a portal that provides movies and web series to users. Here the users can rate a movie that he/she has watched and then based on these ratings can provide these suggestions to other customers who have watched similar movies. Similarly, the target user will also get suggestions for movies that are similar to what he/she has watched or rated highly in the past.

Google News Personalization System

In this case (Arazy et al. 2009), the various news articles are the items and there are no explicit ratings here such. It is a sort of unary feedback. That means that if a user clicks on a particular news item, it is assumed that the user is interested in that news and is taken as positive feedback. So here the feedback mechanism is implicit, and based on this, similar news items will be suggested to the user.

Facebook Friend Recommendations

This social networking site suggests potential friends to users so that there is an increase in the number of social connections on that site. However, the aim of this type of recommendation is slightly different from those of product recommendations. For retailers or merchants, product recommendations increase sales but here there is an increase in social connections. When the number of social connections increases, then this leads to an enhanced experience for a user on that social network. So the company actually depends on this to increase their advertising revenues. So the basis for recommendations for friends or links is actually a link prediction problem in the field of social network analysis. So these systems rely more on structural relationships rather than on the rating data.

In the next section we take a look at the various categories into which recommender algorithms can be classified into.

2.4 Classification of Recommendation Systems

Recommender systems can be broadly classified into three major categories (Ahn et al. 2006; Anand and Mobasher 2005; Balabanovic and Shoham 1997), namely content-based systems, collaborative systems and hybrid systems, as shown in Fig. 2.1.

In the content-based filtering method, similarities in products, services or content features and information gathered about the user are used to make the recommendations.

The advantages of the system are as follows:

- Independent user: There is no need to prepare a similarity index of users for building a personalization system. The recommendations can be done by examining the attributes of the items and the profiles of the users.
- Enough information to avoid cold start: Even if there is very less rating information, new items can be recommended by the others users who are there in the population.
- Transparent Behavior: The method provides the attributes of the items based on which the recommendations have been made.

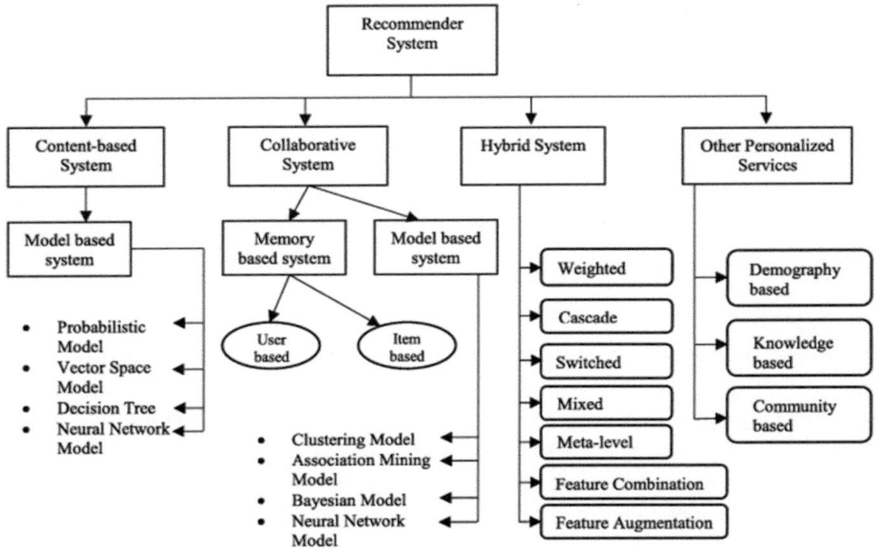

Fig. 2.1 Classification of recommender systems

But it also has certain disadvantages such as:

- Insufficient diversity and novelty: There is a possibility of over specialization which may arise sometimes.
- There is a possibility of inaccurate usage of attributes for the selection of items.
- A lot of domain knowledge is needed if the content based recommender system is to be implemented successfully.
- Bounded content analysis: In case there is not sufficient attribute information, then generating a more accurate recommendation list will be a tough task.
- Sometimes is faced with the "filter bubble" problem where recommendations of items which have already been liked in the past, are made.

In the collaborative filtering method, more weightage is given to find the preferences of similar users and to make recommendations based on them, to another user. The advantages of this system are as follows:

- This system is capable of apprehending changes in user behavior as the time passes.
- It produces a diverse and serendipitous personalization list.
- It provides a solution to the "filter bubble" problem which is faced by the content based systems.
- The collaborative-based system performs really well in large user spaces.
- There is no need for domain information in the initial process of the recommendation.

It also has certain disadvantages like:

- It faces the problem of cold-start item, i.e., if the system hasn't encountered the item in the training phase, then it will be difficult for the system to suggest it in the final personalization list.
- It can turn out to be a complex and expensive system, in cases where there is high dimensionality of data set of very high-dimensional dataset, because calculating the similarity index of millions of users is a tough task for the system.
- Since a majority of the datasets that are available in real life scenarios are sparse, so generating recommendation for such cases using this system may lead the system to recommend in the wrong direction.

In the hybrid systems, there is a condensation of various existing models like content based and collaborative based or any other personalization technique. This was done mainly to overcome the bottlenecks that were being faced by the collaborative system. Its advantages are:

- It is very effective because it combines the benefits of various recommender systems.
- It provides a place for the optimization of the recommendation model.
- The major drawbacks of the content based and collaborative based methods like the cold start problem, the sparsity problem, the gray sheep problem are overcome in this model.

But it has some disadvantages also.

- It is costly to implement it.
- It has high complexity in terms of time and space.
- It uses explicit information, which might pose a problem in data collection due to privacy issues.

A more detailed explanation of the other subcategories is given in later chapters.

2.5 Domain-Specific Challenges in Recommender Systems

In some different domains, temporal data, location-based data, and social data, the context of a recommender system plays a very crucial role (Aimeur et al. 2008; Aimeur and Vezeau 2000).

Context-based recommender data consider many types of information before making a recommendation.

For time-sensitive recommender systems, the ratings of an item may vary with time, because user likes and dislikes evolve with time, e.g., mobile configurations, houses, cars specifications, etc. change frequently over time.

For location-based recommender systems, two types of spatial locality are to be considered—user-specific locality and item-specific locality. So the recommendations for the best places to visit nearby, or places to shop will vary based on the locality that a user is currently in.

Social recommender systems are based on social cues, network structures, tags, or a combination of all three. They are slightly different from the other types of recommendation systems.

2.6 Summary

In this chapter, an overview of recommendation algorithms was given. It discussed the goals of a recommender system and the diverse areas of its applications. It classifies the recommendation algorithms into various categories. The chapter also looks at the challenges that are faced in some specific domains. In the next chapter we take a detailed look into the two most widely used types of recommendation systems, namely collaborative filtering and content-based systems.

Think Tank

1. *What are the goals of a recommender system?*
2. *What are the three main classes of recommender systems? What are their pros and cons?*
3. *What are some of the domain specific challenges in recommender systems?*
4. *Discuss some notable brands who rely heavily on recommendation engines.*

References

Adomavicius G, Sankaranarayanan R, Sen S, Tuzhilin A (2005) Incorporating contextual information in recommender systems using a multidimensional approach. ACM Trans Inf Syst 23(1):103–145

Adomavicius G, Tuzhilin A (2005a) Personalization technologies: a process-oriented perspective. Commun ACM 48(10):83–90

Adomavicius G, Tuzhilin A (2005b) Toward the next generation of recommender systems: a survey of the state-of-the-art and possible extensions. IEEE Trans Knowl Data Eng 17(6):734–749

Ahn H, Kim KJ, Han I (2006) Mobile advertisement recommender system using collaborative filtering: Mar-cf. In: Proceedings of the 2006 Conference of the Korea Society of Management Information Systems, pp 709–715

Aimeur E, Brassard G, Fernandez JM, Onana FSM (2008) Alambic: a privacy-preserving recommender system for electronic commerce. Int J Inf Sec 7(5):307–334

Aimeur E, Vezeau M (2000) Short-term profiling for a case-based reasoning recommendation system. In: de Mantaras RL, Plaza E (eds) Machine learning: 2000, 11th European Conference on Machine Learning. Springer, Berlin, pp 23–30

Anand SS, Mobasher B (2005) Intelligent techniques for web personalization. In: Intelligent Techniques for Web Personalization. Springer, Berlin, pp 1–36

Arazy O, Kumar N, Shapira B (2009) Improving social recommender systems. IT Prof 11(4):38–44

Averjanova O, Ricci F, Nguyen QN (2008) Map-based interaction with a conversational mobile recommender system. In: The Second International Conference on Mobile Ubiquitous Computing, Systems, Services and Technologies. UBICOMM'08, pp 212–218

Baccigalupo C, Plaza E (2006) Case-based sequential ordering of songs for playlist recommendation. In: Roth-Berghofer T, Goker MH, Guvenir HA (eds) ECCBR, vol 4106. Lecture notes in computer science. Springer, Berlin, pp 286–300

Bailey RA (2008) Design of comparative experiments. Cambridge University Press, Cambridge

Balabanovic M, Shoham Y (1997) Content-based, collaborative recommendation. Commun ACM 40(3):66–72

Chapter 3
Collaborative Filtering and Content-Based Systems

Abstract In this chapter, the two most widely used types of recommender systems, namely the collaborative filtering method and the content-based system, along with a few of their important sub-types are discussed in this chapter. There are two types of collaborative methods, namely the neighborhood-based and model-based methods. The chapter discusses what are the features of and differences between the two methods. The basic components of the content-based systems are also discussed. Both the systems have their advantages and disadvantages which are also discussed here.

Keywords Content based · Collaborative filtering · Similarity measures · Ratings · Recall · Precision · RMSE

3.1 Introduction

The previous chapter gave an overview of the various types of algorithms used in recommendation systems. Since the two broad categories of recommendation algorithms are the collaborative filtering model and the content-based recommender systems; therefore this chapter explains the two methods in more detail. In CBS, the ratings and the buying patterns of users are combined with the descriptive attributes of the items to arrive at the recommendations. Here the descriptions of the item are given some ratings and fed as training data for the creation of a regression model to classify users.

So the system stores the descriptions of all the items bought or rated by a particular user, and this information is used to predict whether a user will be interested in a new product or not. In the collaborative filtering method, the system depends on previous interactions between users and items to generate new suggestions. So it classifies the groups of similar users into clusters and then provides suggestions to the users based on the preferences of the people in the clusters. So the idea is that users in one cluster are likely to have similar preferences or choices of products or movies, etc.

P. Kar et al., *Recommender Systems: Algorithms and their Applications*, Transactions on Computer Systems and Networks, https://doi.org/10.1007/978-981-97-0538-2_3

We discuss the details of the models in the next sections (Aslanian et al. 2016; Beel et al. 2013; Bellogin et al. 2011).

3.2 Collaborative Filtering Model

Collaborative filtering methods (Fig. 3.1) (Bobadilla et al. 2011; Bogers and Bosch 2008; Das et al. 2007) can be broadly classified into two categories: neighborhood-based collaborative filtering and model-based collaborative filtering. So we describe each of them separately below:

- *Neighborhood-based collaborative filtering*—They are also called **memory-based algorithms** and are among the earliest algorithms that have been used for CF. It is based on the assumption that users with similar profiles have similar patterns in which they rate items and that similar items get similar ratings. It is of two types—*user-based CF and item-based CF* (Silva et al. 2017b; Su and Khoshgoftaar 2009). In user-based CF methods, the system stores the ratings of products

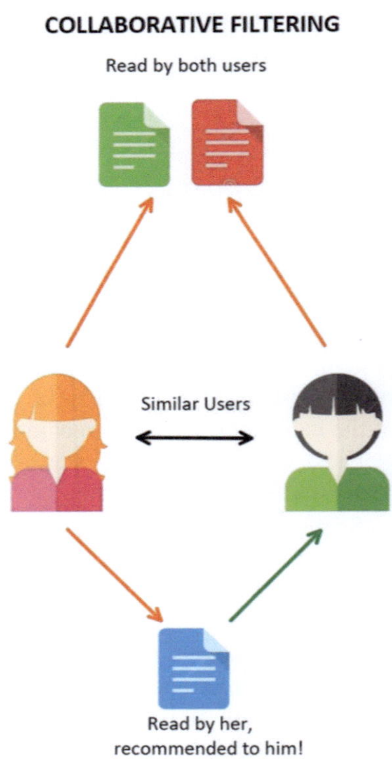

Fig. 3.1 Collaborative filtering method

by users who have profiles similar to the target user and then suggests these products to the target user. In item-based CF, a set of items with features that are similar to a target item are selected first. Then the weighted average of all the ratings of a particular user is used to predict the ratings of a user for that item. The main difference between user-based and item-based CF is that in the first case the ratings are predicted using the ratings of the neighboring users, whereas in the second case the predictions are based on the ratings of a user on neighboring items. The algorithms for this method can be formulated in any one of the following ways: either by predicting the rating value of a use-item combination or by determining the top k items or top *k* users. The ratings of the matrices in a rating matrix can be of four types: continuous ratings, interval-based ratings, ordinal ratings, binary ratings, and unary ratings. In continuous ratings, there is a continuous scale that corresponds to the degree of liking or dislike of an item. In interval-based ratings, the ratings are usually on a 10-point or 20-point scale. Ordinal ratings, it's similar to interval-based ratings but some predefined categories are available, e.g., agree, accept, weak accept, strongly accept, reject, neutral, etc. In binary ratings, as the names suggest, only two options are available, yes or no. In unary ratings, a user is allowed to specify only a positive preference but not a negative preference, i.e., the like button on Facebook.

There are mainly two principles that are there in neighborhood models: user-based models and item-based models (Goldberg et al. 1992; Herlocker et al. 2004). In the user-based model's users with similar profiles usually give similar ratings on the same items. So ratings of one user for an item may be used as a basis for recommending it to other users with similar profiles. In the item-based model, if some items are similar, then a user is likely to give similar ratings to those items.

- *Model-based collaborative filtering*—Here the recommendations for the items are provided by first developing a model of the ratings of the user. So here the algorithms use a probabilistic approach and use the collaborative filtering method to compute the expected value of the user prediction, based on the ratings given by a user on other items. The following are the advantages of the model-based system over the neighborhood-based system.

Space efficiency—The size of the learning model is usually a lot smaller than the original rating matrix. So the space requirements for the model-based system are low, whereas the item-based or user-based methods have higher complexity in the order of $O(n^2)$.

Training speed and prediction speed—The preprocessing is faster in the model-based systems as compared to the neighborhood models as it is quadratic in the number of users or number of items in the case of the latter system. In a majority of the cases, the compact and summarized model can efficiently make predictions.

Avoiding overfitting—A lot of machine learning algorithms suffer from the problem of overfitting. This means that random artifacts in the data try to overly influence the process of prediction. This issue is also there in classification and random models. To avoid the problem of overfitting a summarization approach may be used. In addition, adding regularization methods can also be used for making robust models.

3.3 Content-Based Recommender Systems

The collaborative method mentioned in the previous section is based on correlations for their ratings patterns to arrive at their recommendations. These methods don't use item attributes to make predictions. In the content-based approach (Noia et al. 2012; Fernandes et al. 2017), there is more focus on items that can be described by some descriptive set of attributes. So basically the ratings of a user itself on other products are sufficient to make accurate recommendations. This type of method is found to be useful in situations where there are fewer ratings for an item. Content-based recommender systems (Fig. 3.2), attempt to match users to products that are similar to what products the user has already liked in the past. The similarity may not be based on the correlations of the ratings, but they can also be based on the attributes of the items which a user likes. So while collaborative filtering uses the ratings of other users and that of the target, in the content-based systems, the ratings and the attributes of the products liked by the target user are used. So other users do not have much of a role to play in the content-based systems, as it uses a different source of data for the recommendations. The two sources of data are as follows:

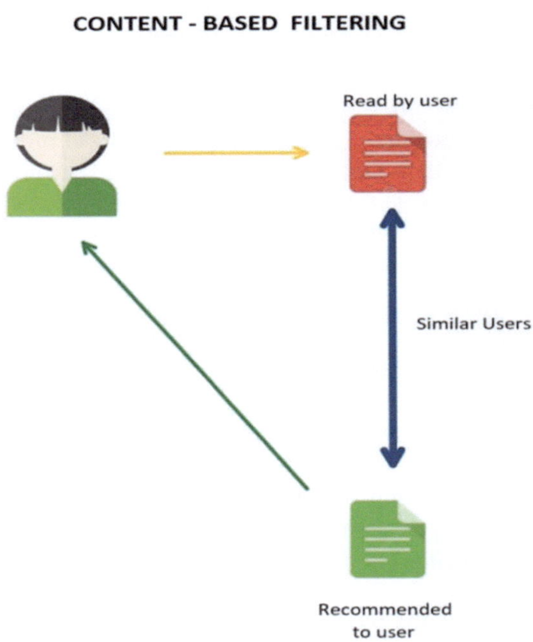

Fig. 3.2 Content-based filtering

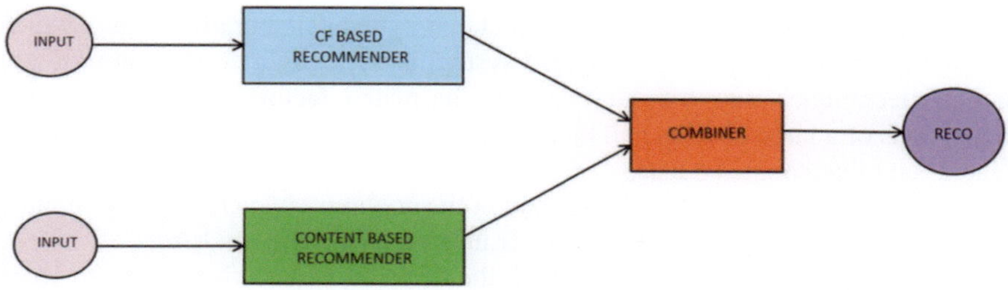

Fig. 3.3 Hybrid systems

The first source of data is a description of different items as content-centric attributes.

The second source of data is the profile of a user, which is based on user feedback for different products. The feedback of the user can be either explicit or implicit. The process of collection of the ratings is similar to that of the collaborative systems. The basic components of the content-based system are as follows:

Pre-processing and feature extraction—A wide variety of domains use content-based systems like web pages, news, music, etc. So here the features are extracted from the different sources and are changed to keyword-based vector space representations. For any content-based recommender system, this is the first step and is very domain specific. But extracting the features which give the maximum information is crucial for the effective working of a content-based recommender system.

Content-based learning of user profiles—A content-based model is specific for a particular user as we have seen earlier. So the models which are built for the prediction of user interest for products based on their history of either buying or rating a product are user-specific. For this, the feedback of a user is utilized and this can be explicit feedback or implicit feedback. These feedbacks are used along with the item attributes for the construction of the training data. Based on this training data, a learning model is built. This step is similar to classification or regression modeling, depending on the fact that the feedback is categorical (i.e., binary process of selection of an item) or numerical (i.e., there are frequencies of ratings or buyings). The resultant model is called the user profile as it relates the user interests to the attributes of objects.

Filtering and Recommendation—Here the model that was learned in the previous step is utilized for making suggestions for products for specific users. This step needs to be very efficient as the predictions have to be done in a real-time scenario.

3.4 Hybrid Recommender Systems

The hybrid approach (Burke 2002; Chowdhury 2010; Glauber et al. 2013) is basically a combination of various existing models, like the content based, collaborative based or any of the personalization techniques (Fig. 3.3). This method came up a solution to overcome the bottlenecks that were encountered by the collaborative system which was the most used system. So a hybrid system is basically a combination of one or more techniques, e.g., a system can use the matrix factorization to reduce the dimensions of a large data set and then collaborative filtering can be applied for the generation of personalized lists.

The pros and cons have been described in the previous chapter. So at this point a definition of the various hybrid methods (Liu et al. 2018) is given here.

In the weighted hybridization method, the decision is made based on the score obtained from different recommender systems. Then the results of each of the recommender systems is collated to a single numerical component for deciding the final recommendation list.

In the cascade method of hybridization, the basis of the recommendation is a chain of recommendations, which means that the results of one recommendation system are fine-tuned based on the results of another recommendation system.

In the switched method of hybridization, the method chooses a suitable recommendation system from the set of the recommendation systems.

In the mixed hybridization method, as the name suggests, different recommendation techniques work together to create a collaborative decision on the final personalized list.

In the meta level method, the output from one recommendation system is taken as the input for another recommendation system.

In the feature combination method, various types of knowledge source features are aggregated together to form a single domain.

In the feature augmentation method, the features of one knowledge source is calculated so that it becomes compatible to become the input for some other recommendation algorithm.

Apart from the three major approaches described above, there are also various other personalized services, like demography based systems, knowledge-based systems and community-based systems (Wu et al. 2015).

In the demography-based system, the users are categorized based on the demographic data like gender, age, qualification, location, etc. This type of system is difficult to implement in real-life scenarios, because it is very difficult to gather correct and complete demographic data of users.

In the knowledge based system, the recommendations are based on the needs of the user. So it uses the knowledge about a user and the item to decide which if the items will fulfill the user needs. So it depends on services based on user preferences (Konstan and Riedl 2012; Lathia et al. 2010).

In the community-based systems, communities are made by the recommendation system based on people who share common interests. It relies on a user-item interaction within a community, and recommendations of items are made after an aggregate decision is obtained from the community.

3.5 Similarity Measures Used by a Recommender System

How well a recommender system performs, depends entirely on similarity measures (McNee et al. 2006; Said and Bellogín 2014; Santana et al. 2017; Shani and Gunawardana 2011). So the accuracy of the model depends on the preciseness of the similarity calculation. All the methods have their own pros and cons but the Pearson Correlation Coefficient (PCC) is the most widely used similarity measure for recommendation systems. The similarity measures can be either:

- Distance-based similarity
- Correlation-based similarity

3.5.1 Distance-Based Similarity Measure

This can be measured in any one of the following three ways:

i. Euclidean distance—This method measures the distance between any two points z_1 and z_2 using the formula:

$$\text{dist}(z_1, z_2) = \text{dist}(z_2, z_1) = \sqrt{\sum_{i=1}^{n}(z_1 - z_2)^2} \tag{3.1}$$

where z_1 and z_2 are any two points or objects in Euclidean space whose similarity needs to be evaluated.

ii. Manhattan distance—This method computes the distance on gridlines. The calculation is done by summing the vertical and horizontal component of any set of points. So the Manhattan distance between any two points z_1 and z_2 is given by the following formula:

$$\text{dist}(z_1, z_2) = \text{dist}(z_2, z_1) = \sum_{i=1}^{n}|z_1 - z_2| \tag{3.2}$$

iii. Minkowski distance—This is a generalized representation of both the Manhattan distance and the Euclidean distance. The Minkowski distance between any two points z_1 and z_2 can be any set of real values as shown in the formula below:

$$\text{dist}(z_1, z_2) = \text{dist}(z_2, z_1) = \left(\sum_{i=1}^{n}|z_1 - z_2|^p\right)^{\frac{1}{p}} \tag{3.3}$$

3.5.2 Correlation-Based Similarity

This can also be calculated in the following four ways:

i. Pearson Correlation Coefficient—In this method, the similarity is calculated between objects or points based on common items or ratings. The similarity value varies between − 1 and + 1, where − 1 represents negative correlation and + 1 represents positive correlation and 0 represents no correlation. The Pearson correlation value between two points z_1 and z_2 is calculated according to the following formula:

$$\mathrm{PC}(z_1, z_2) = \frac{\sum_{i=1}^{n}\left(z_{1i} - z_2'\right)\left(z_{2i} - z_2'\right)}{\sqrt{\sum_{i=1}^{n}\left(z_{1i} - z_1'\right)^2 \sum_{i=1}^{n}\left(z_{2i} - z_2'\right)^2}} \tag{3.4}$$

ii. Cosine similarity—This method is mostly used for high-dimensionality positive spaces.

It is used to measure the similarity between any two objects based on certain attributes. The cosine similarity between an object z_1 and a point z_2 can be calculated according to the following formula:

$$\cos(\theta) = \frac{\sum_{i=1}^{n} z_{1i} \cdot z_{2i}}{\sqrt{\sum_{i=1}^{n} z_{1i}^2 \sum_{i=1}^{n} z_{2i}^2}} \tag{3.5}$$

iii. Adjusted cosine similarity—This method of cosine similarity takes into consideration the changing rating scale of the users. So the adjusted cosine similarity between any two items I_1 and I_2 can be calculated with the following formula:

$$\text{Adj. Cos. } (I_1, I_2) = \frac{\sum_{u \in U_{I_1, I_2}}\left(r_{uI_1} - r_u'\right)\left(r_{uI_2} - r_u'\right)}{\sqrt{\sum_{u \in U_{I_1, I_2}}\left(r_{uI_1} - r_u'\right)^2}\sqrt{\sum_{u \in U_{I_1, I_2}}\left(r_{uI_2} - r_u'\right)^2}} \tag{3.6}$$

where U_{I_1, I_2} represents the set of users who have rated both items I_1 and I_2, and r_{uI_1} and r_{uI_2} denote the ratings that have been given by user 1 to I_1 and I_2 respectively, and r_u' is the average rating give by user u.

iv. Jaccard similarity—It is an index value that is used for calculating the similarity and diversity between a set of objects. It is defined as division of intersection over union. The similarity value varies between 0 and 1, where 0 represents low similarity and 1 represents high similarity. The Jaccard similarity between any two objects z_1 and z_2 is given by the following formula:

3.6 Evaluation Metrics of Recommender Systems (Ge et al. 2010)

3.6.1 Mean Absolute Error (MAE)

This method is a measurement of the deviation of the predicted value from the actual value and is calculated according to the following formula:

$$\text{MAE} = \frac{\sum_{i=1}^{n} |A_i - P_i|}{n} \tag{3.7}$$

3.6.2 Root Mean Square Error (RMSE)

This is used for the calculation of error during the prediction of value of an object and is the square root of the difference between the predicted and the actual values and is calculated according to the following formula:

$$\text{RMSE} = \sqrt{\frac{\sum_{i=1}^{n} (A_i - P_i)^2}{n}} \tag{3.8}$$

where A_i and P_i are the actual and predicted values respectively and n represents the total number of items for which predictions have been made.

3.6.3 Precision

This is defined as the total number of relevant items in the recommendation list divided by the total number of items in that list and is calculated by the following formula:

$$\text{Precision}(P) = \frac{\text{True Positive}}{\text{True Positive} + \text{False Positive}} \tag{3.9}$$

where true positive is the number of items which are relevant and present in the list and false positive is the number of items which are not relevant but still present in the list.

3.6.4 Recall

This is defined as the ratio of the relevant items in the recommendation list divided by the total number of relevant items in the population and is calculated by the following formula:

$$\text{Recall}(R) = \frac{\text{True Positive}}{\text{True Positive} + \text{False Negative}} \tag{3.10}$$

3.6.5 F_1 Score and Accuracy

F_1 score is the harmonic mean between precision and recall with the following formula:

$$F_1 \text{ Score} = \frac{2 * P * R}{(P + R)} \tag{3.11}$$

Accuracy is the total number of correct prediction divided by the population of items present in the system and is calculated as:

$$\text{Accuracy} = \frac{\text{True Positive} + \text{True Negative}}{\text{True Positive} + \text{True Negative} + \text{False Positive} + \text{False Negative}} \tag{3.12}$$

3.7 Summary

So in this chapter, the concepts of two widely used systems, namely collaborative filtering and content-based recommendation systems were introduced (Shardanand and Maes 1995; Silva et al. 2017a). This chapter explains the two main types of collaborative filtering methods and their features. The next method is the content-based method which differs from the collaborative filtering method in that it depends on the past ratings and likings of similar items by the target user itself instead of collecting the ratings of other users with similar profiles for those items. It also describes the hybrid methods and some personalized methods. After that the methods for calculating the similarity measures have been discussed, and followed by the various evaluation metrics of a recommendation system. The next chapter discusses the matrix decomposition and clustering process.

Think Tank

1. *What is a content based recommendation system?*
2. *What is a collaborative filtering case recommender system?*
3. *What are the types of hybrid recommendation system?*
4. *What are the evaluation metrics of a recommender system?*
5. *What are the similarity measures used by recommendation systems?*

References

Aslanian E, Radmanesh M, Jalili M (2016) Hybrid recommender systems based on content feature relationship. IEEE Transactions on Industrial Informatics

Beel J, Genzmehr M, Langer S, Nürnberger A, Gipp B (2013) A comparative analysis of offline and online evaluations and discussion of research paper recommender system evaluation.In: Proceedings of the international workshop on reproducibility and replication in recommender systems evaluation. ACM, pp 7–14

Bellogin A, Castells P, Cantador I (2011) Precision-oriented evaluation of recommender systems: an algorithmic comparison. In: Proceedings of RECSYS. ACM, pp 333–336

Bobadilla J, Ortega F, Hernando A, Alcalá J (2011) Improving collaborative filtering recommender system results and performance using genetic algorithms. Knowl Based Syst 24(8):1310–1316

Bogers T, Van den Bosch A (2008) Recommending scientific articles using citeulike. In: Proceedings of RECSYS. ACM, pp 287–290

Burke R (2002) Hybrid recommender systems: survey and experiments. User Model User-Adapt Interact 12(4):331–370

Chowdhury G (2010) Introduction to modern information retrieval. Facet Publishing, Abingdon

Das AS, Datar M, Garg A, Rajaram S (2007) Google news personalization: scalable online collaborative filtering. In: Proceedings of 7 WWW. ACM, pp 271–280

Di Noia T, Mirizzi R, Ostuni VC, Romito D, Zanker M (2012.) Linked open data to support content-based recommender systems. In: Proceedings of Semantics. ACM, pp 1–8

Fernandes BB, Sacenti JA, Willrich R (2017) Using implicit feedback for neighbors selection: alleviating the sparsity problem in collaborative recommendation systems. In: Proceedings of WEBMEDIA. ACM, pp 341–348

Ge M, Delgado-Battenfeld C, Jannach D (2010) Beyond accuracy: evaluating recommender systems by coverage and serendipity. In: Proceedings of RECSYS. ACM, pp 257–260

Glauber R, Loula A, Rocha-Junior JB (2013) A mixed hybrid recommender system for given names. ECML PKDD Discov Challenge 2013:25–36

Goldberg D, Nichols D, Oki BM, Terry D (1992) Using collaborative filtering to weave an information tapestry. Commun ACM 35(12):61–70

Herlocker JL, Konstan JA, Terveen LG, Riedl JT (2004) Evaluating collaborative filtering recommender systems. ACM Trans Inf Syst 22(1):5–53

Konstan J, Riedl J (2012) Recommender systems: from algorithms to user experience. User Model User-Adapt Interact 22(1):101–123

Lathia N, Hailes S, Capra L, Amatriain X (2010) Temporal diversity in recommender systems. In: Proceedings of ACM SIGIR. ACM, pp 210–217

Liu Y, Wang S, Khan MS, He J (2018) A novel deep hybrid recommender system based on auto-encoder with neural collaborative filtering. Big Data Mining and Analytics 1(3):211–221

McNee SM, Riedl J, Konstan JA (2006) Being accurate is not enough: how accuracy metrics have hurt recommender systems. In: Proceedings of CHI. ACM, pp 1097–1101
Said A, Bellogín A (2014) Comparative recommender system evaluation: benchmarking recommendation frameworks. In: Proceedings of RECSYS. ACM, pp 129–136
Santana LL, Souza AB, Santana DL, Dourado WA, Durão FA (2017) Evaluating ensemble strategies for recommender systems under metadata reduction. In: Proceedings of WEBMEDIA. ACM, pp 125–132
Shani G, Gunawardana A (2011) Evaluating recommendation systems. In: Recommender systems handbook. Springer, Berlin, pp 257–297
Shardanand U, Maes P (1995) Social information filtering: algorithms for automating "word of mouth". In: Proceedings of SIGCHI. ACM Press/Addison-Wesley Publishing Co., Boston, MA, pp 210–217
Silva DV, Silva RD, Durão FA (2017a) RecStore: recommending stores for shopping mall customers. In: Proceedings of WEBMEDIA. ACM, pp 117–124
Silva N, Carvalho D, Pereira AC, Mourão F, Rocha L (2017b) Evaluating different strategies to mitigate the ramp-up problem in recommendation domains. In: Proceedings of WEBMEDIA. ACM, pp 333–340
Su X, Khoshgoftaar TM (2009) A survey of collaborative filtering techniques. Adv Artif Intell 2009:4
Wu D, Zhang G, Lu J (2015) A fuzzy preference treebased recommender system for personalized business-to-business e-services. IEEE Trans Fuzzy Syst 23(1):29–43

Chapter 4
Matrix Decomposition for Clustering and Collaborative Filtering

Abstract Since the consumers of today are flooded with choices for various products like movies on OTT platforms, online music, and other online shopping sites, so to increase user satisfaction and maintain loyalty, the retailers and content providers need to find ways to match users with their most preferred products of choice. So they use recommender systems which have been very successful in providing accurate suggestions of items to customers. The two main strategies used by recommender systems are content-based models and collaborative filtering. Matrix factorization is a collaborative filtering method to find the relationship between items' and users' entities. Latent features, the association between users and movies matrices, are determined to find similarity and make a prediction based on both item and user entities. Matrix factorization is a way to generate latent features when multiplying two different kinds of entities. Since not every user gives ratings to all the items they use, there are many missing values in the matrix and it results in a **sparse matrix.** Hence, the null values not given by the users would be filled with 0 such that the filled values are provided for the multiplication. It has been observed that matrix factorization models are superior when compared to the nearest neighbor technique, for the generation of product recommendations. This is because it incorporates additional factors like implicit feedback, temporal effects, and confidence levels into the recommendation process. Therefore in this chapter, we see how the process of matrix decomposition works, in detail.

Keywords Matrix decomposition · Clustering · Sparse matrix · User preferences · Matrix factorization

4.1 Introduction

At a time when consumers are faced with a huge variety of options of products to choose from, the success of the retailers depends on how accurately they can predict user preferences and choices, and suggest new products which the user has a very high probability of buying. Recommendations can be generated by a wide

P. Kar et al., *Recommender Systems: Algorithms and their Applications*, Transactions on Computer Systems and Networks, https://doi.org/10.1007/978-981-97-0538-2_4

variety of algorithms. As we have already seen, the two main strategies used by recommender systems are collaborative filtering and content-based methods. The user-based or item-based collaborative filtering methods are simple and intuitive, but matrix factorization techniques are generally more effective because they allow the discovery of the latent features underlying the interactions between users and items. The two main areas of collaborative filtering are the neighborhood methods and the latent factor models. The neighborhood model computes the relationships between items or between users. The first category finds the preferences of a user for a particular item, by finding the ratings of "neighboring" items by the same user. A neighbor of a product means that those products are likely to get similar ratings when reviewed by the same user. For example, if we consider the movie "Black Hawk Down", then neighbors of that particular movie will include movies like, those which involve wars or are directed by Ridley Scott. So if we want to predict the rating that a particular user has given to the movie "Black Hawk Down", then we need to search for the ratings that the user has given to similar movies that the user has watched.

Another alternative approach that tries to predict ratings by characterizing items and users through inference of some rating patterns is latent factor models. These factors are a computerized alternative to the options created by humans (Sarwar et al. 2000; Funk 2006). For example, if we consider movies, there can be several additional factors that can be considered, like comedy vs drama, the amount of action, children's orientation, and other such dimensions which are usually not very well defined. Latent factor models rely on matrix decomposition to discover latent features to find the underlying interactions between any two types of entities. So in the next section, an overview of the matrix decomposition process is given (Koren 2008; Paterek 2007; Takács et al. 2007).

4.2 Why Matrix Decomposition is Useful?

Matrix decomposition/factorization is a mathematical technique that can be applied to scenarios where we would like to find some hidden components in data. Matrix factorization /decomposition is a process by which two matrices are factorized so that their product gives the original matrix. Netflix or other such movie recommendation systems consist of a set of viewers/users and a set of movies. Now assuming that each of the users has given ratings to some movies, we need to predict the ratings of the users for the movies that they have not yet rated so that the users can get proper recommendations. So if we have say 5 users and 5 items, then we can represent all the rating information in the form of a 5×5 matrix, as shown in Fig. 4.1. The ratings will be integers ranging from 1 to 5 and if there is a hyphen if a user has not yet rated a movie. So now the problem at hand is to fill in the blanks/hyphens so that the values given as ratings are consistent with the ratings which are already there in the matrix.

The reason behind using matrix factorization for solving the above-mentioned problem is that there must be some hidden features that will determine how an item

	D1	D2	D3	D4	D5
U1	5	3	-	1	3
U2	4	-	-	1	2
U3	1	4	3	-	-
U4	-	3	2	4	-
U5	-	-	5	3	4

Fig. 4.1 Matrix for user ratings for movies with blanks/hyphens

is rated by a user. So suppose if two users give high ratings for a particular movie, then there might be a common factor for both of them liking the movie like maybe both of them like the same actor/actress, or both of them like action movies as their preferred genre of movies. If we are able to find these hidden features, then it would be easier to predict the ratings of a particular user about a particular item, as we can match the features of the user with that of the item. So when we try to uncover these features, we assume that the number of these features is less than the number of users and items. The assumption is valid since otherwise, it would mean that each user has a unique feature, which although not entirely impossible, is a rare phenomenon. Secondly, if this situation really happened, then making recommendations would be fairly useless, as the other users would not be interested in the movies that have been rated by other users((Salakhutdinov and Mnih 2008; Bell and Koren 2007)).

4.3 The Matrix Decomposition Technique

In this section, we take a look at the mathematical process of matrix factorization. Suppose there is a set of **U** users and **D** items. R is a matrix of size $|U| \times |D|$ containing all the ratings given by the users to the items. The assumption is finding **K** latent features. So now the job is to find two matrices P ($|U| * K$) and Q ($|D| * K$), such that their product approximates to R:

$$R \approx P \times Q^T = \hat{R} \tag{4.1}$$

Here the rows of **P** represent the strengths of the associations between a user and the features, whereas the rows of Q represent the strengths of the associations between an item and the features. So to predict the rating of an item d_j by u_i, the dot product of the vectors which correspond to u_i and d_j, is calculated.

$$\hat{r}_{ij} = p_i^T q_j = \sum_{k=1}^{k} p_{ik} q_{kj} \tag{4.2}$$

The objective is to find a method to find **P** and **Q.** One way of doing this is to first initialize the two matrices with some values and then calculate how their product differs from M, after which the difference is minimized iteratively. This process is known as gradient descent and is aimed to find a local minimum of the difference. The difference is also called the error between the estimated rating and the real rating and can be calculated by the equation shown below for each pair of user-item:

$$e_{ij}^2 = (r_{ij} - \hat{\boldsymbol{r}}_{ij})^2 = \left(r_{ij} - \sum_{k=1}^{K} \boldsymbol{p}_{ik}\boldsymbol{q}_{kj}\right)^2 \tag{4.3}$$

Here the squared error is considered as the estimated rating can be higher or lower than the real rating. For error minimization, it is necessary to know the direction in which the values of p_{ik} and q_{kj} have to be modified, i.e., the gradient at the current values needs to be known. So the above equation is differentiated with respect to these variables separately:

$$\frac{\partial}{\partial_{p_{ik}}} e_{ij}^2 = -2(r_{ij} - \hat{\boldsymbol{r}}_{ij})(\boldsymbol{q}_{kj}) = -2e_{ij}q_{kj} \tag{4.4}$$

$$\frac{\partial}{\partial_{q_{kj}}} e_{ij}^2 = -2(r_{ij} - \hat{\boldsymbol{r}}_{ij})(\boldsymbol{p}_{ik}) = -2e_{ij}p_{ik} \tag{4.5}$$

After getting the gradient, the update rules for both p_{ik} and q_{kj} can be formulated as follows:

$$\boldsymbol{p}_{ik}^{'} = \boldsymbol{p}_{ik} + \boldsymbol{\alpha}\frac{\partial}{\partial_{p_{ik}}} e_{ij}^2 = \boldsymbol{p}_{ik} + 2\boldsymbol{\alpha} e_{ij}q_{kj} \tag{4.6}$$

$$\boldsymbol{q}_{kj}^{'} = \boldsymbol{q}_{kj} + \boldsymbol{\alpha}\frac{\partial}{\partial_{\boldsymbol{q}_{kj}}} e_{ij}^2 = \boldsymbol{q}_{kj} + 2\boldsymbol{\alpha} e_{ij}p_{ik} \tag{4.7}$$

Here α is a constant and its value gives the rate of approaching the minimum and is normally taken to be a small value like 0.0002. The reason behind this is that if the steps toward the minimum are taken to be too large, then there is a chance of missing the minimum which will lead to oscillations around the minimum. The update rules are applied to iteratively perform the operations until the error converged to its minimum. The overall error can be checked by the equation given below, which tells when to stop the process:

$$E = \sum_{(u_i, d_{i,} r_{ij}) \in T} e_{ij} = \sum_{(u_i, d_{i,} r_{ij}) \in T} \left(r_{ij} - \sum_{k=1}^{K} \boldsymbol{p}_{ik}\boldsymbol{q}_{kj}\right)^2 \tag{4.8}$$

The matrix that is obtained from implementing the above algorithm is as follows:

As we see here, the approximations obtained are very close to the actual ratings, and some predictions about the unknown values can also be made.

4.4 The Netflix Example

The online movie rental company Netflix announced a contest in 2006 for improving their recommender system. So the company released a training set consisting of more than 100 million ratings spanning over 500,000 anonymous customers and how they rated more than 17,000 movies, where each movie was rated on a scale of 1–5 stars, for the teams to work on. The teams which took part submitted the predicted ratings for a test set of approximately 3 million ratings. Netflix calculated the RMSE (root-mean-square error) based on the held-out truth. The challenge was that whichever team was the first to improve on Netflix's algorithm's RMSE performance by 10% or more would win prize money of $1 million.

If none of the teams succeeded in reaching the 10 percent goal, then a prize of $50,000 was given to the team in the first place, meaning with the least RMSE, after every yearly competition. This competition generated a lot of interest in the field of collaborative filtering because, until that point in time, the data that was publicly available for research in collaborative filtering was many magnitudes smaller than what was released by Netflix. So the release of this data created a flurry of activities and research worldwide. As per the Netflix website, there were more than 48,000 teams from 182 counties who had downloaded the data.

A group consisting of Yehuda Koren from Yahoo Research and Robert Bell and Chris Volinsky from AT&T Labs—Research won the top spot in 2007 with their entry names BellKor and won the Progress Prize for 2007 with what was the best score at that time: 8.43% better than Netflix, and later joined with the team Big Chaos to win the 2008 Progress Prize with 9.46% improvement.

The winning entries had greater than 100 different predictor sets, most of which were factorization models or variants of the model discussed above. When the Netflix user-movie matrix is factorized, it gives the most descriptive parameters for the prediction of user preferences for movies. The first two factors from the Netflix data matrix factorization are shown in Fig. 4.2 where movies are placed based on their factor vectors. The first-factor vector on the x-axis has comedies on one side and horror movies targeted at male or teenage audiences, and the other side contains serious undertones and strong female leads. The second-factor vector on the y-axis has independent, critically acclaimed, weird films, and on the bottom mainstream formula films. There are various films on the intersection of these films. Thus it was seen that matrix factorization is a very crucial method in collaborative filtering methods. By using it successfully to the Netflix Prize data, it has been found that they offer much better accuracy as compared to the nearest neighbor technique. The property that makes them even more convenient is that these models can naturally integrate many important aspects of the data, like multiple feedback forms, temporal data, and confidence levels (Figs. 4.3, 4.4).

```
import numpy

def matrix_factorization(R, P, Q, K, steps=5000, alpha=0.0002, beta=0.02):
    Q = Q.T
    for step in xrange(steps):
     for i in xrange(len(R)):
      for j in xrange(len(R[i])):
        if R[i][j] > 0:
          eij = R[i][j] - numpy.dot(P[i,:],Q[:,j])
            for k in xrange(K):
              P[i][k] = P[i][k]+alpha*(2 * eij * Q[k][j] - beta* P[i][k])
                Q[k][j] = Q[k][j] + alpha * (2 * eij * P[i][k]-beta* Q[k][j])

        eR = numpy.dot(P,Q)
        e = 0
        for i in xrange(len(R)):
          for j in xrange(len(R[i])):
            if R[i][j] > 0:
              e = e + pow(R[i][j] - numpy.dot(P[i,:],Q[:,j]), 2)
               for k in xrange(K):
                 e = e + (beta/2) * (pow(P[i][k],2) + pow(Q[k][j],2))
        if e < 0.001:
            break
    return P, Q.T
```

(a)

```
R = [
     [5,3,0,1],
     [4,0,0,1],
     [1,1,0,5],
     [1,0,0,4],
     [0,1,5,4],
    ]

R = numpy.array(R)

N = len(R)
M = len(R[0])
K = 2

P = numpy.random.rand(N,K)
Q = numpy.random.rand(M,K)

 nP, nQ = matrix_factorization(R, P, Q, K)
nR = numpy.dot(nP, nQ.T)
```

(b)

Fig. 4.2 **a** Code snippets for matrix factorization in Python. **b** Code snippet to be run on the above algorithm, containing many zero values

Fig. 4.3 Resultant matrix

	D1	D2	D3	D4
U1	4.97	2.98	2.18	0.98
U2	3.97	2.40	1.97	0.99
U3	1.02	0.93	5.32	4.93
U4	1.00	0.85	4.59	3.93
U5	1.36	1.07	4.89	4.12

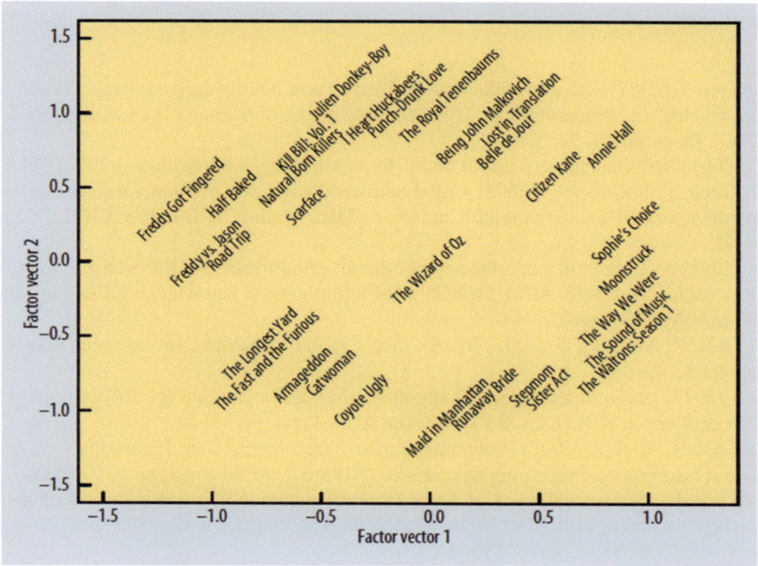

Fig. 4.4 First two vectors from a matrix decomposition of the Netflix Prize data. The selected movies were placed at the appropriate spot based on their factor vectors in two dimensions (Zhou et al. 2008; Koren et al. 2009)

4.5 Summary

In this chapter, we gave an overview of the two main methods of recommendation systems, namely collaborative filtering, and content-based methods (Hu et al. 2008). Of the two methods in collaborative filtering, the latent factor models have been found to give better recommendation results as compared to the neighborhood model. This is because it uses matrix factorization to find out latent factors for the ratings of a user. The mathematical process of matrix factorization has also been discussed here. Finally, an overview of the famous Netflix challenge, which uses the matrix factorization method is given here. Till now we were looking at recommendations as a rating prediction problem, but an interesting and alternative approach would be to see if it can now be treated as a ranking problem i.e. how to arrange, display or stack the results in some particular order. In the next chapter we discuss the process of learning how to rank and how to collect user behavior.

Think Tank

1. *What is a sparse matrix?*
2. *What is matrix factorization?*
3. *Give examples of latent features in user matrices.*
4. *What are the reasons which lead to a rating matrix with null values?*
5. *What is the advantage of using matrix factorization in recommendation systems?*

References

Bell R, Koren Y (2007) Scalable collaborative filtering with jointly derived neighborhood interpolation weights. In: Proceedings on IEEE International Conferencce Data Mining (ICDM 07). IEEE CS Press, pp 43–52

Funk S (2006) Netflix update: try this at home. http://sifter.org/~simon/journal/20061211.html

Hu YF, Koren Y, Volinsky C (2008) Collaborative filtering for implicit feedback datasets. In: Proceedings on IEEE International Conference Data Mining (ICDM 08). IEEE CS Press, pp 263–272

Koren Y (2008) Factorization meets the neighborhood: a multifaceted collaborative filtering model. In: Proceedings on 14th ACM SIGKDD Int'l Conference Knowledge Discovery and Data Mining. ACM Press, pp 426–434

Koren Y, Bell R, Volinsky C (2009) Matrix factorization techniques for recommender systems. Computer 42:42–49

Paterek A (2007) Improving regularized singular value decomposition for collaborative filtering. In: Proceedings on KDD Cup and Workshop. ACM Press, pp 39–42

Salakhutdinov R, Mnih A (2008) Probabilistic matrix factorization. In: Proceedings on Advances in Neural Information Processing Systems 20 (NIPS 07). ACM Press, pp 1257–1264

Sarwar BM et al. (2000) Application of dimensionality reduction in recommender system—a case study. In: Proceedings on KDD Workshop on Web Mining for e-Commerce: Challenges and Opportunities (WebKDD). ACM Press

Takács G et al (2007) Major components of the gravity recommendation system. SIGKDD Explor 9:80–84

Zhou Y et al. (2008) Large-scale parallel collaborative filtering for the Netflix prize. In: Proceedings on 4th International Conference Algorithmic Aspects in Information and Management, LNCS 5034. Springer, pp 337–348

Chapter 5
Learning How to Rank and Collecting User Behavior

Abstract Till now we were looking at recommendations as a rating prediction problem, but an interesting and alternative approach would be to see if it can now be treated as a ranking problem, i.e., how to arrange, display or stack the results in some particular order. So sometimes it may make more sense if we stack the most relevant choice at the top, followed by the second most relevant choice at the second place, and so on. In this chapter, we take a look at the ranking methods used by FourSquare and Facebook, as well as some LtR (learning to rank algorithms). Many of these algorithms have been first used in IR (Information Retrieval), and many of these algorithms have also been used quite successfully without much hindrance. While on the topic of ranking, we also need to be careful about fake user profiles who deliberately give biased feedbacks to increase or decrease the rank of an object. So here we also take a look at how to collect the likes and dislikes of a user and how to filter fake profiles.

Keywords LtR algorithms · Rank · Filter · Fake user profile · Biased feedback · Forward selection · Backward selection · Shilling attack · Obfuscated attack · Detection algorithms

5.1 Introduction

Most of the techniques discussed in the previous chapters treat the recommendation problem as a prediction problem and rarely present all the ratings to the users. The system usually suggests the top n items to the user. Moreover, a user normally pays more attention to the results at the top of the list as compared to the items which are ranked lower. So some predicted values may not be displayed to the user or optimized predicted values may not always provide the best recommendations. The main reason for this is that the objective functions of prediction-based methods are not fully aligned with the experience of the end user (Adomavicius and Tuzhilin 2005; Hu et al. 2008).

P. Kar et al., *Recommender Systems: Algorithms and their Applications*, Transactions on Computer Systems and Networks, https://doi.org/10.1007/978-981-97-0538-2_5

"Learning to rank" combines different types of data sources like distance, popularity, or the outputs from a recommender system. The main thing here is that the rank does not always need to be a part of the recommender system. So basically, during ranking one looks at the input sources that can help in the ordering of the objects. In this chapter, we take the example of a popular app named FourSquare to show how the learning-to-rank technique is used for ordering suggestions. After that, we take a look at the various types of LtR algorithms (Koren and Sill 2011; Liu 2009; Radlinski et al. 2008). After that, we see the process of how to detect a fake profile and a review of the various types of attacks on a recommender system and how to detect them.

5.2 The Facebook and FourSquare (Factual) Ranking Methods

FourSquare (now Factual) has a website and a mobile app that is basically a guide to cities. It is a city search platform, which provides personalized recommendations to users about nearby places based on the location, to find the best matches for nearby places to visit, the best places to shop, or the best restaurants based on the preferences of that particular user. A person can search for information and reviews on various places and events in a particular geographical area. It also learns the preferences of a user over time and can predict and suggest the places the person is likely to go to, even when the person is visiting another place anywhere in the world. The user can search for information and reviews about various facilities and events in any part of the world.

Suppose a person is in a new place and is searching for the list of nearby coffee shops around him/her. The person will get a list of recommendations of coffee shops around him/her pushed into the phone or in the web application. Usually, these suggestions may not always be arranged in the order of nearest distances or ratings of the restaurants. So how does the system work in FourSquare? It uses some features like spatial score, timeliness, popularity, here now, personal history of the user, etc. to arrive at the ratings. But they do not include parameters like distance or ratings of the places. In order to incorporate these and obtain the revised rankings, one needs to design an appropriate weighted function and train the machine learning algorithm with it. So basically the machine is now trained to rank the places based on distance and ratings to get the optimized rankings. So a simplified view of the FourSquare problem is shown in Fig. 5.1. Other relevant parameters may also be chosen to make the ranking optimized. In the next section, we take a look at some LtR(Learning to Rank) algorithms. One thing that needs to be kept in mind here is that while a hybrid recommender system predicts ratings, the LtR algorithms produce the orderings.

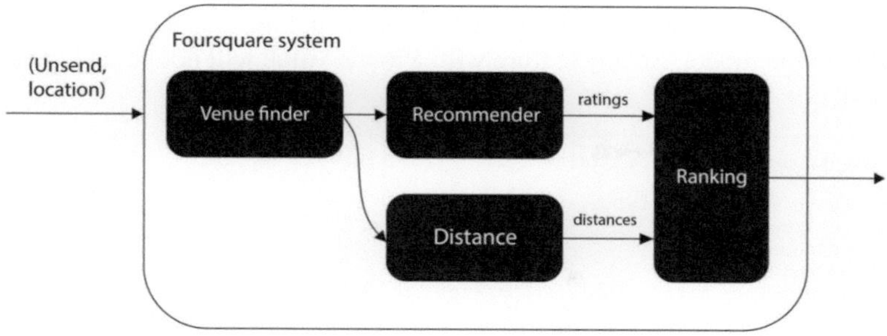

Fig. 5.1 A simplified view of the Foursquare app

5.3 Feature Selection in Recommender Systems

Feature selection (Manning et al. 2008) is a very important process in the designing of efficient recommendation algorithms. So what is a feature? A feature is basically an X-variable in the datasets and is usually defined by a column. Nowadays datasets can have more than 100 + features, which make them very difficult to work with normally. So in such cases feature selection techniques come in very handy. What feature selection does is to reduce the number of features that are included in a model but without sacrificing the predictive power of the model. So we normally look for features that are redundant, irrelevant can actually affect the performance of the model negatively, so it will be useful if can identify these features and remove them.

So the main benefit of feature selection is that it prevents overfitting, by removing extraneous data and helps the model to focus on the important aspects of the data and helps in increasing the accuracy of the predictions made by the model. There are three types of feature selection methods: **Wrapper methods** (forward, backward, and stepwise selection), **Filter methods** (ANOVA, Pearson correlation, variance thresholding), and **Embedded methods** (Lasso, Ridge, Decision Tree).

- **Wrapper methods**—These models start with a particular subset of features and calculates the importance of each of those features. Then the model iterates and tries other different subsets of features, until an optimal subset is found. The challenge with this method is that for datasets with very large number of features it will require a very high computation time, and because of this it will overfit the model when the number of data points is less. The important wrapper methods for feature selection are forward selection, backward selection, and stepwise selection.

The **forward selection** methods initially start with zero features, and for each of these features it will run a model and find the p-value related to the t-test or F-test performed on them. Then a selection of the feature with the lowest p-value is made and added to the working model. Then it will take the first feature and run the models with a second feature added to it and select the second feature with the lowest p-value. Similarly, it will take two features previously selected and run the model with a third feature and so on. Therefore only those features which have a significantly *p*-vales are added to the model. So any feature with a low *p*-value will not be included in the final model.

```
def forward_selection(X, y, initial_list=[], threshold_in=0.01,
verbose=True):
    included = list(initial_list)
    while True:
        changed=False
        # forward step
        excluded = list(set(X.columns)-set(included))
        new_pval = pd.Series(index=excluded)
        for new_column in excluded:
            model = sm.OLS(y,
sm.add_constant(pd.DataFrame(X[included+[new_column]]))).fit()
            new_pval[new_column] = model.pvalues[new_column]
            return included
```

In the **backward selection process**, it starts with all the features that are there in the dataset and then runs the model to calculate the *p* value associated with the *t*-test or the *F*-test for each feature. The feature which has the largest insignificant *p*-value will be removed from the model and then the process is started again iteratively until all the features which have insignificant *p*-values are removed from the model.

```
def back_selection(X, y, threshold_out = 0.05,verbose=True):
 included = X.columns.tolist()
 while True:
     model = sm.OLS(y, sm.add_constant(pd.DataFrame(X[included]))).fit()
     pvalues = model.pvalues.iloc[1:]
     worst_pval = pvalues.max() # null if pvalues is empty
     if worst_pval > threshold_out:
         worst_feature = pvalues.argmax()
         included.remove(worst_feature)
         return included
```

The **stepwise selection method** is a hybrid of the forward and backward selection methods. Here, the process starts with 0 features and adds the feature with the lowest significant *p*-value. Then it will find the second feature with the lowest significant *p*-value. In the third iteration, it finds the next feature with the lowest significant *p*-value and also remove any of the features that were previously added but at present have insignificant *p*-values.

```
def stepwise_selection(X, y,initial_list=[],threshold_in=0.01,threshold_out =
0.05,verbose=True):
    included = list(initial_list)
    while True:
        excluded = list(set(X.columns)-set(included))
        new_pval = pd.Series(index=excluded)
        for new_column in excluded:
            model = sm.OLS(y,
sm.add_constant(pd.DataFrame(X[included+[new_column]]))).fit()
            new_pval[new_column] = model.pvalues[new_column]
        best_pval = new_pval.min()
        model = sm.OLS(y, sm.add_constant(pd.DataFrame(X[included]))).fit()
        #check the entire function on my github page
        if not changed:
            break
        model = sm.OLS(y, sm.add_constant(pd.DataFrame(X[included]))).fit()
    return included
```

- **Filter Methods**—Filter methods do not use error rate as a parameter to determine if a feature is useful or not. Here a subset of the features is selected by ranking them by using a useful descriptive measure. The advantage of this method is that it has low computation time and does not overfit the data. But one disadvantage is that it is not capable of detecting any interactions or correlations between the features. The three different methods here are ANOVA, Pearson correlation, and variance thresholding.

The ANOVA (Analysis of Variance) process, as the name says, sees the variation within the treatments of a feature as well as in between treatments. These variances are useful parameters for this particular method because here we can find out if a feature is properly accounting for variations in the dependent variable.

```
def ANOVA(X,y):

    '''Univariate linear regression tests

    Quick linear model for sequentially testing the effect of many regressors

    Using scikit learn's Feature selection toolbox

    Returns:

        F (array) = F-values for regressors

        pvalues (array) = p-values for F-scores'''

    (F,pvalues) = f_regression(X,y)

    return (F,pvalues)
```

The **Pearson correlation coefficient** measures the similarity of two features that range between − 1 and 1. If any two features have a value close to − 1 or 1 then it implies that the two features may be related or have a high correlation to each other. The cutoff value of a high correlation vs low correlation is dependent on the range of the correlation coefficients in a dataset.

The third method in this category is **variance thresholding**. The variance of a feature gives its predictive power. So the lower the variance of a feature, the lesser will be the information that will be contained in the feature and will therefore have less value in the prediction of the response variable.

So based on this fact, variance thresholding is done where the variance of each feature is found and then all of features are dropped below a particular variance threshold. If the threshold is 0 then it will only remove the features that have the same value for each case of the response variable. If more features are to be removed, then the threshold could be set to a higher value.

```
# Create VarianceThreshold object with a variance with a default threshold of
0.5

def variance_threshold_selector(data, threshold=0.5):

    selector = VarianceThreshold(threshold)

    selector.fit(data)

    return data[data.columns[selector.get_support(indices=True)]]
```

- **Embedded Methods**—These methods use a part of the model creation process for the process of feature selection. This puts these methods at a convenient medium in between the two previously discussed methods as these methods do the selection along with the model tuning process. Lasso and Ridge regression are the two most widely used feature selection method in this category, and Decision tree also creates a model using a different technique for feature selection.

In statistics, overfitting is the production of an analysis that corresponds too closely or exactly to a particular set of data, and may therefore fail to fit additional data or predict future observations reliably. So an overfitted model is a statistical model that has more parameters than what can be justified by the data. The most obvious consequence of overfitting is poor performance on the validation dataset.

If we want to keep all the features in the final model, but at the same time do not wish that the model focuses too much on any one single coefficient, then **Ridge regression** achieves this by putting a penalty on the beta coefficients of a model if it is too large. It is a regularization method to reduce overfitting. A trend line that overfits the training data is used and therefore it has a much higher variance than the Ordinary Least Squares (OLS) method. The basic logic behind Ridge regression is to fit a new line which doesn't fit the training data. It means we introduce a certain amount of bias into the new trendline. The inability of a machine learning algorithm to capture the true relationship is called Bias. In Machine Learning, an algorithm ideally needs to have Low Bias and should be able to accurately approximate the true relationship. In practice, a bias called Lambda is introduced and the penalty function is ***lambda*slope^2***. The lambda is a penalty term and this value is called ridge regression or L2. The L2 penalty is quadratic. When lambda is zero the penalty

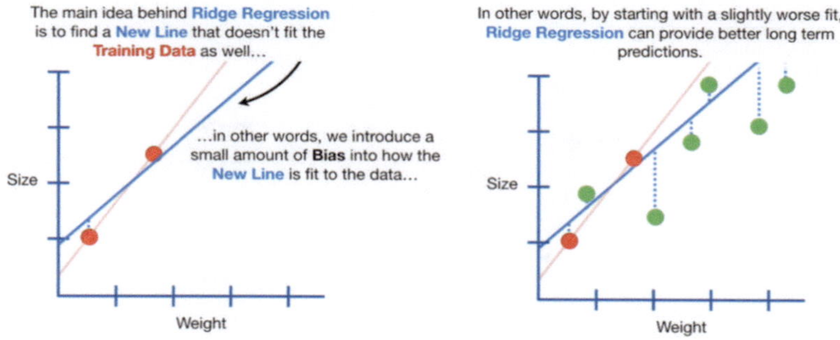

Fig. 5.2 Graph of ridge regression with bias

is also zero, and therefore only the sum of squared residuals is minimized. When lambda increases asymptotically, we arrive at a slope which is close to zero, so the larger is the value of lambda, the less sensitive the prediction becomes to the independent variable (Fig. 5.2).

The **Lasso Regression** is also a regularization method to reduce overfitting. It also puts a penalty on the beta coefficients in a model, but it also adds a penalty term to the cost function of the model with a lambda value that has to be tuned. It is similar to the Ridge regression but with only one very major difference: the penalty function is now lambda*|slope|. The main difference between the Ridge regression and the Lasso Regression is that the Lasso Regression can force the beta coefficient to zero so that the feature is removed from the model. So if we are looking to reduce the model complexity, Lasso regression is the preferred method. The result of the Lasso Regression is also similar to the result given by the Ridge regression. Both of them can be used for Logistic regression, regression with discrete values, and regression with interaction. The difference between the two methods can be realized when we increase the value on lambda. Ridge can only shrink the slope asymptotically close to zero, whereas Lasso can shrink the slope all the way to zero (Fig. 5.3). The advantage is evident when we lots of parameters in the model. In Ridge if we increase the value of lambda, the most important parameters may shrink a little, and the less important parameters stay with a high value. In contrast, in Lasso when the value of lambda is increased, the most important parameters shrink little, but the less important parameters get near to zero. In this way Lasso can exclude the unimportant parameters from the model.

The **Decision Tree** is another method for feature selection and uses a regression tree or a classification tree depending on whether the response variable is continuous or discrete respectively. Decision tree Regressor builds a tree incrementally by splitting the dataset into subsets which give rise to a tree with decision nodes and leaf nodes. A decision node has two or more branches where each value represents the attribute tested. The Leaf node represents the decision on the numerical target. The topmost node is called the root node which corresponds to the best predictor.

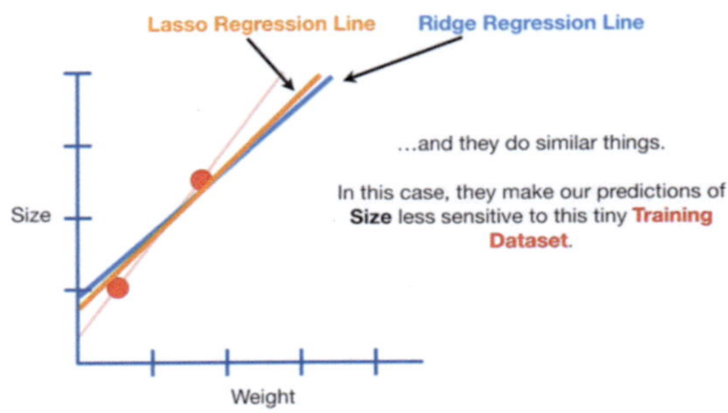

Fig. 5.3 Lasso regression in comparison with ridge regression

It works by creating splits in the tree depending on certain features for the creation of the algorithm to find the response variable. At each split, the function that was used to create the tree will check all possible splits for all the features and will choose the feature that splits the data into the most homogeneous groups. It basically means that it selects the feature that predicts best, what the response variable will be at each point in the tree.

5.4 The Ranking Module in a Recommender System

A typical recommendation system that is based on matching and ranking, normally consists of two modules, the matching module and the ranking module (Radlinski et al. 2008; Rendle et al. 2009). When a user visits a site, he/she will find a wide number of articles from which to select a particular item. So to provide a better and faster experience to the user, it is necessary to filter out the items that the user may be interested in. The job of the matching module is to provide a preliminary filtering and reduce the size of the item list that is to be delivered to the ranking module. So it basically means that the matching module will first select a small proportion of items that the user may be interested in, from a large collection of items. So that means that if the site has say, 100,000 items to offer, then the matching module will filter out maybe around only 500 of those items that the user may be interested in, from the full list of items. E.g. if the matching module has shortlisted say 500 items from the list, then it means that the user may like these 500 items, but it is not known what is the order or rank of preference of those items for the user. So now to rank these 500 items, in the order of the users most favorite to the least favorite, we need to apply an appropriate ranking algorithm that is based on the user properties and the item properties. The ranking algorithm will therefore now rank these 500 items based on the preferences of the user and create a final list of items which are ranked from most favorite to least favorite, and sent to the user. So in the recommendation system, the matching module does the initial filtering of items to reduce the list of items that the ranking module has to rank. In this way, the property-based ranking of items by the ranking module is sped up and leads to a more efficient recommendation feedback mechanism. A professional recommendation service needs to be capable of providing recommendation feedback within only a few milliseconds of receiving a request from a user. Refreshing the feed stream should take no more than a few milliseconds, after which the newly recommended item list must be pushed to the user.

5.5 Introduction to Ranking Algorithms

The LtR algorithms (Learning to Rank) can be broadly classified into three categories based on their ways of evaluating the ranked list during the training phase—pointwise, pairwise, and listwise, as shown in Fig. 5.4.

- *Pointwise*—In this system, a score is produced for each item and then they are ranked accordingly. This is similar to the approaches in the recommendation systems in previous chapters. Rating prediction is different from ranking in that ranking does not care about the utility score of an item being even one million, as long as the score is a valid rank in the system.
- *Pairwise*—As the name suggests, this is a binary classifier that uses a function to take two items as the input and returns the ordering of the two items. So they are basically pairs of items where a user has given the preference. The pair only has the information on whether the first item is preferred to the second item or not by giving a $+1$ or a -1. In this binary classification problem, the aim is to optimize the output so that the learning method implicitly aims to minimize the number of pairwise inversions in the training data.
- *Listwise*—This is the best LtR approach as it takes the entire ranked list and optimizes it. Listwise ranking is preferred because it understands that ordering is more important at the top of a list as compared to the bottom of the list. The pointwise and pairwise algorithms can't differentiate where an item is on the ranked list.

But with the gradual advent of deep learning, the algorithms for ranking are being gradually integrated with deep learning. There are four types of algorithms which are typically used for ranking now and are as follows: the logistic regression (LR), the

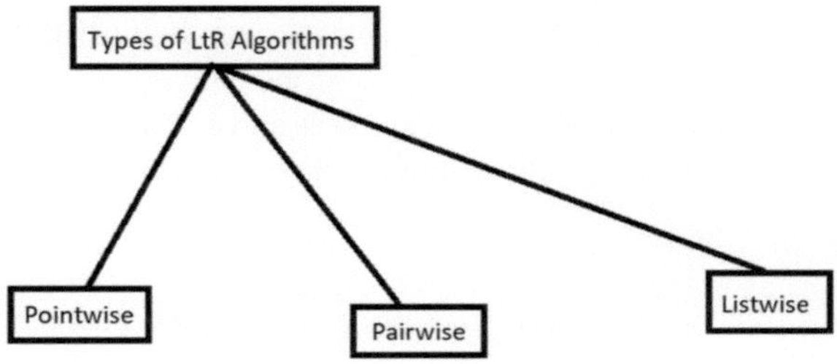

Fig. 5.4 Types of LtR algorithms

factorization method (FM), gradient boosted decision trees(GBDT), and the DeepFM models.

The logistic regression model, (LR) is the most classic binary algorithm and is easy to use and needs low computation power. The factorization machine (FM) has been applied in the past few years on various customer scenarios and has been found to give promising results and uses the inner product method for feature representation. The third method is a logistic regression method that uses gradient boosting decision trees (GBDTs) and feature encoding for increasing the interpretability of the data features. The fourth algorithm is the DeepFM algorithm and uses a combination of deep learning and classis learning algorithms.

5.6 Collecting User Likes and Dislikes

Apart from collecting content about the items for a recommendation, it is also very important to obtain information about the likes and dislikes of a particular user, in order to complete the process of recommendation. While the collection of data is done during the offline phase, the recommendations are found during the online phase, when a particular user is interacting with a system. An active user is someone for whom the prediction is done at any given point in time. So during the online phase, the personal preferences of the user are combined with the content for the creation of predictions. The data related to the likes and dislikes can be in any one of the following forms:

Ratings—Here the users will specify ratings that will indicate their preferences for a particular item. The ratings may be binary, interval based, ordinal, or even real-valued. The choice of the rating type usually has a big impact on the type of model that is to be used for learning about the user profiles.

Implicit feedback—This defines the user-actions like buying or browsing an item. In a majority of the cases, only the positive preferences of the user are collected along with implicit feedback, but negative preferences are not collected.

Text Opinions—In some instances, the opinions expressed by the user might be in the form of text descriptions also. If that happens, then the implicit ratings can be extracted from these opinions. This type of extraction of rating deals with opinion mining and sentiment analysis.

Cases—Users sometimes may also specific examples or cases of items that he/she is interested in. Then these cases can be utilized as implicit feedback for some algorithms.

In all of the above cases, the likes and dislikes of a user about an item are converted to unary, binary, interval-based, or real ratings.

5.7 Detection of Fake/Malicious Profiles

With the present situation of information overload, consumers have a tough time finding and selecting the information that is most relevant to them. Recommender systems are a boon for potential customers because they use information filtering techniques to help the consumers to make their choices of items to select. However, with the rise of users and items featured on the recommender systems, new challenges have also come up. The collaborative filtering technique is one of the widely used recommendation systems in use at present. But unfortunately, it is also extremely prone to shilling/profile injection attacks. The effect of these attacks is that they adversely affect the recommendation process to promote or demote a particular attack. In this section, we give an overview of the various types of shilling attacks and some detection algorithms (Bhaumik et al. 2006, 2007). A more comprehensive study is outside the scope of this book.

As we have already discussed in Chap. 3, recommendation systems can be broadly classified into two types, i.e., collaborative filtering-based and content-based. The content-based approach works by recommending the products to the users by comparison of the products to the profiles of the users.

The collaborative filtering recommender system on the other hand analyses the last behavior of a user to find the best matches. It is based on the assumption that users with similar behaviors will have similar interests. So it basically depends on the relationship between the users and the items. But unfortunately, because of its openness and dependency on user ratings, collaborative filtering is very often prone to shilling attacks or profile injection attacks.

A shilling attack is a type of attack, where a malicious user profile is deliberately inserted into an existing collaborative filtering data set so that the outcome of the recommendation system gets changed. The result is that these injected profiles will explicitly rate the items in such a manner that the main target item will either get promoted or demoted.

We explain the effect of a shilling attack with the following example:

Suppose there are only two users A and B in a system, and they have given similar ratings to some products say p1, p2, and p4. Now if user B gives a high rating to product p3, then p3 will also be recommended to user A. So basically it finds the top x users who are similar to the target user u and then the ratings of the products for the user u are calculated based on similar users' ratings of the products, and the top few products with high ratings, which have not yet been rated by user u, are then recommended to the user u. So whenever a new user with a similar profile gives a high rating to a particular product, then that product will be recommended to the other users with similar profiles. In the same way, if some new users give a low rating to a particular product, then the chances of that product being recommended to other users with similar profiles become low.

5.8 Shilling/Profile Injection Attacks

In Fig. 5.5, product X gets promoted and recommended to other users based on the high ratings of a malicious user who has injected his profile into the recommendation system. Shilling attacks can be classified into two categories, a push attack or a nuke attack, depending on what purpose it is being used for. If it is being used to gain promotion for an item then it is a push attack and if it is being used to demote an item then it is termed a nuke attack, and both types are used to gain an edge or profit over a competitor.

There are different types of shilling attacks as shown in Fig. 5.6, and they are broadly classified as standard attacks and obfuscated attacks (Bryan et al. 2008; Burke et al. 2006). Standard attacks do not make a special attempt to not get detected in a recommender system. So the detection algorithms have a higher chance of detecting these types of shilling/profile injection attacks. Some examples of this type of attack are random attack, average attack, bandwagon attack, reverse bandwagon attack, segmented attack, probe attack, and love/hate attack. Obfuscated attacks on the other hand try to prevent themselves from getting detected by obfuscating their attack signatures. Most of these methods make small modifications to the standard techniques to obtain obfuscation. The obfuscation may reduce the impact of the attack sometimes but the plus side is that they have lesser chances of getting detected. Some techniques of this type are noise injection, user shifting, target shifting, average over popular, mixed attack, power item attack, power item attack, SAShA.

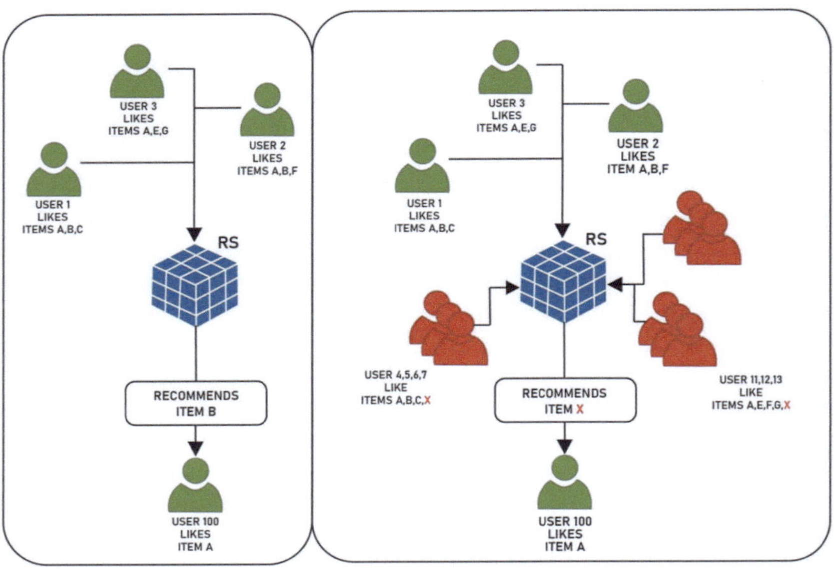

Fig. 5.5 An example of a Shilling attack

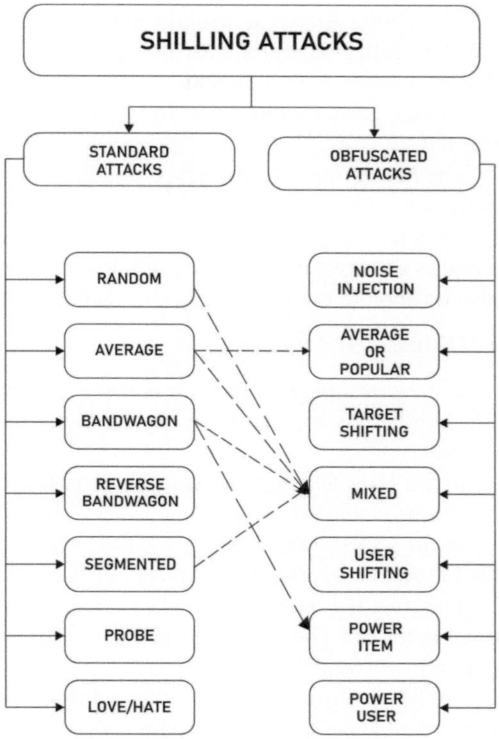

Fig. 5.6 Types of shilling attacks

- ***Standard Attacks***

Random Attack or the RandomBot attack is the simplest type of shilling attack where the items rated by the attack profile are chosen randomly, except the target item. The ratings for all these items are around the system overall mean, and the target item is given the maximum or minimum rating depending upon whether it's a push attack or a nuke attack. It is easy to implement but not very effective. It's used more to disrupt a recommendation system than to actually promote an item.

The *Average Attack* is similar to the random attack as far as selecting the item is considered.

In the *Bandwagon Attack*, the profiles that are generated by the attackers are filled with popular items with high ratings, and the target item is given the highest ratings.

The *Reverse Bandwagon Attack* is the opposite of the Bandwagon Attack where the target product is given the lowest ratings and is used for nuke attacks.

In the *Segmented Attack*, a specific group of users who are likely to buy an item in an e-commerce setup are targeted. This attack has a high impact as it is aimed at a particular segment.

The *Probe Attack* is usually not a generalized one for all systems. Here the attacker utilizes the predicted rating scores that are projected by some recommendation systems and gives genuine ratings to some items. When the recommendation system suggests further items, the attacker can make the rated list of items based on this list.

The *Love/Hate Attack* is a type of nuke attack which is very effective. In this filler, items are chosen randomly by the attacker and given the highest ratings, while the target items are given the lowest ratings. Although it seems like a simple model, it's a highly effective model. Although it was basically designed for nuke attacks, it can also be applied to push attacks, but then they will not be as effective in this case.

- **Obfuscated Attacks**

The *Noise Injection* method adds a Gaussian distributed random number which is multiplied by a constant, to each of the ratings, which are a subset of the infected profiles.

In the *User Shifting* technique, a subset of the rated item of each of the injected profiles is changed.

In *Target Shifting*, the ratings of the target items are shifted to one level lesser than the highest that is possible in push attacks.

The *Average over Popular* technique obfuscates the Average attacks where the filler items are chosen from among the top x% of the items which are the most popular, with equal probability.

The *Mixed Attack* is achieved by applying the random, average, bandwagon, and segmented attacks in equal proportions.

The *Power Item Attack* uses the power items which are chosen by some particular methods, where power items are defined as a set of items that is capable of influencing the largest number of items.

The *Power User Attack* is similar to the above attack, but here the set of users who have the maximum influence on the broadest group of users is chosen.

SAShA is a technique that uses the semantic features which are extracted from a knowledge graph for improving the standards of the usual Collaborative filtering attack models.

5.9 Detection Algorithms

There are various types of detection algorithms and they can be broadly classified as supervised and unsupervised detection methods. Since the supervised techniques need the data to be labeled during the training process, and labeled data is very less in recommendation systems, therefore the unsupervised methods are used more here.

A majority of these detection algorithms target a particular trait in a shilling attack. Even though obfuscation makes it possible to evade detection to some degree, there must be some innate features present in the attack to make it effective to a certain extent. The traits can be user-based traits or item-based traits (Fig. 5.7).

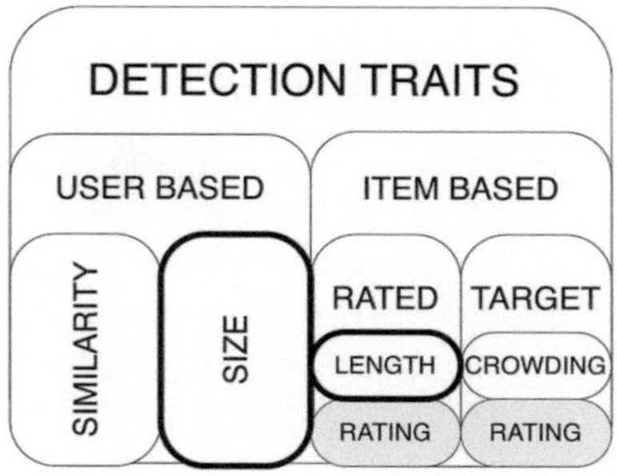

Fig. 5.7 Trait-based detection algorithms for shilling attacks

User-based traits include the similarity of a profile to a large number of user profiles and the size of the attack, i.e., the number of injected attack profiles.

Item-based traits are the length and ratings of the *rated items*, the number of users, and the ratings of the *target items*.

Along with the various algorithms for the detection of shilling attacks, a parallel line of research focuses on the creation of robust algorithms that are immune to shilling attacks. Although these algorithms do not use a technique to remove find and remove shilling profiles, they can reduce the effect of shilling attacks.

5.10 Summary

In this chapter, we have discussed a very crucial topic of how to properly rank the items in a recommendation system and an overview of the types of ranking algorithms. The chapter also discusses the problems of fake profiles and how the presence of malicious users can adversely affect the outcomes of suggestions by a recommendation system and reduce its credibility. An overview of the shilling attack and its types has also been given here along with techniques to detect such types of attacks and the corresponding actions to be taken. In Chap. 8, we have discussed how to build trust centric and attack resistant recommendation systems. The recommendation algorithms that have been described in the previous chapters used past ratings of users, and preferences of users with similar profiles to arrive at their suggestions of products for new users. But in some types of situation if sufficient past ratings are not available, or there are complex variations in the combinations of preferences of some particular products, then the aforementioned methods are not very effective. This is especially true for products that are not bought frequently or high-end luxury products which have high levels of personal customizations or the preferences of users evolve over time. To handle such situations, knowledge-based and hybrid, and ensemble-based techniques are highly useful to give accurate suggestions. These are discussed in the next chapter.

Think Tank

1. *What is a ranking problem?*
2. *What are the ranking methods used for Facebook and FourSquare apps?*
3. *What are the forward selection and backward selection methods?*
4. *What are the various types of LtR algorithms?*
5. *In what forms can we store the data related to user likes and dislikes?*
6. *What is a Shilling attack and what are its types?*
7. *What are the available detection algorithms for Shilling attacks?*

References

Adomavicius G, Tuzhilin A (2005) Toward the next generation of recommender systems: a survey of the state-of-the-art and possible extensions. IEEE Trans Knowl Data Eng 17(6):734–749

Bhaumik R, Williams C, Mobasher B, Burke R (2006) Securing collaborative filtering against malicious attacks through anomaly detection. In: Workshop on Intelligent Techniques for Web Personalization (ITWP)

Bhaumik R, Burke R, Mobasher B (2007) Crawling attacks against web-based recommender systems. In: International Conference on Data Mining (DMIN), pp 183–189

Bryan K, O'Mahony M, Cunningham P (2008) Unsupervised retrieval of attack profiles in collaborative recommender systems. In: ACM Conference on Recommender Systems, pp 155–162

Burke R, Mobasher B, Williams C, Bhaumik R (2006) Classification features for attack detection in collaborative recommender systems. In: ACM KDD Conference, pp 542–547

Hu Y, Koren Y, Volinsky C (2008) Collaborative filtering for implicit feedback datasets. In: ICDM '08. IEEE Computer Society, pp 263–272

Koren Y, Ordrec JS (2011) An ordinal model for predicting personalized item rating distributions. In: Proceedings of the fifth ACM conference on Recommender systems, RecSys'11. ACM, pp 117–124

Liu T-Y (2009) Learning to rank for information retrieval. Found Trends Inf Retr 3(3):225–331

Manning CD, Raghavan P, Schutze H (2008) Introduction to information retrieval, 1st edition. Cambridge University Press, Cambridge

Radlinski F, Kleinberg R, Joachims T. Learning diverse rankings with multi-armed bandits. In: Proceedings of the 25th international conference on Machine learning, ICML'08. ACM, New York, NY, pp 784–791

Rendle S, Freudenthaler C, Gantner Z, Lars S-T (2009) Bpr: Bayesian personalized ranking from implicit feedback. In: UAI'09. AUAI Press, pp 452–461

Chapter 6
Knowledge-Based, Ensemble-Based, and Hybrid Recommender Systems

Abstract The recommendation algorithms described in the previous chapters used past ratings of users, and preferences of users with similar profiles to arrive at their suggestions of products for new users. But in situations where sufficient past ratings are not available, or there are complex variations in the combinations of preferences of some particular products, the previous methods are not very effective. This is especially true for products that are not bought frequently or high-end luxury products which have high levels of personal customizations or the preferences of users evolve over time. This causes the cold start problem which is a well-known challenge in recommendation systems. In such cases to avoid the cold start problem, knowledge-based and hybrid, and ensemble-based techniques are highly useful to give accurate suggestions.

Keywords Knowledge based system · Ensemble based systems · Hybrid systems · Cold start problem · Customization

6.1 Introduction

Sometimes users require information on some categories of items that are not bought very frequently. Some examples of such products are houses, cars, tourism plans, financial services, or some costly luxury products. So there may not be a sufficient amount of ratings available for the recommendation process to work on (Aggarwal et al. 2001). Since the items are purchased less frequently and have various types of detailing and combinations, it is difficult to get sufficient ratings for a particular combination of parameters for a product. The cold start problem is a similar problem that is encountered in recommendation systems when sufficient ratings are not available to it. In addition to this, the preferences of the consumers may also change over time. For example, the preferences for the models and specifications of cars may change over time, so the past data stored about previous user ratings may not be useful anymore. Also, there may be different combinations of parameters that are relevant to different users like color, engine capacity, fuel efficiency, brand, interior

P. Kar et al., *Recommender Systems: Algorithms and their Applications*, Transactions on Computer Systems and Networks, https://doi.org/10.1007/978-981-97-0538-2_6

options, etc. So basically it's a very complex task to generate sufficient ratings for such a varied combination of parameters. These types of cases are dealt with by using knowledge-based recommender systems, which do not use ratings as a basis for their recommendation process. Here the recommendation is based on the similarities between the requirements of the users and the descriptions of the products, or the constraints that are specified by the user. So the process basically uses a knowledge base and hence the name of this approach. So while content-based systems or collaborative filtering are based entirely on the past actions or ratings of a user, or the action or ratings of people with similar profiles, knowledge-based systems are different in the way that they tell the users to explicitly specify their requirements (Bobadilla et al. 2013; Bohnert et al. 2008).

Sometimes when the inputs have a wide variety, then a person has the flexibility to use various types of recommender systems for performing the same task. This gives rise to the option of hybridization where a combination of different types of systems is implemented to get the best of both. Hybrid recommender systems have a close relation to the area of ensemble analysis, where multiple machine learning algorithms are combined together for the creation of a more robust model. Ensemble-based recommender systems not only use multiple data sources but also increase the effectiveness of a particular type of recommender system by implementing the combination of multiple models of the same type. In this chapter the knowledge-based, ensemble-based, and hybrid recommendation systems are discussed. A brief introduction of the cold start problem is also given here since it is an interesting problem and also a popular topic for research.

6.2 The Cold Start Problem

The cold start problem is very well known in recommendation systems and is also an interesting aspect for research. This problem is usually faced in computer-based information systems, where some amount of data modeling is involved. To be more precise, it deals with the problem where the system cannot draw any conclusions or infer for the users or items about which the system has not yet gathered sufficient information. Since recommender systems involve information filtering, which enables it to try to present information of items that may be of possible interest to the user. To do this, a recommendation system compares the profile of a user to some reference characteristics. These characteristics may be characteristics of the item or the past behavior of a user. So based on what the system is designed to do, the user can be linked to different types of interactions like ratings, purchases, number of page visits, etc. There are three cases of the cold start problem: *new community, new item and new user* (Fig. 6.1).

In the new community case, also known as systemic bootstrapping, is when the system is at the point of startup, when there is almost no user interaction and no

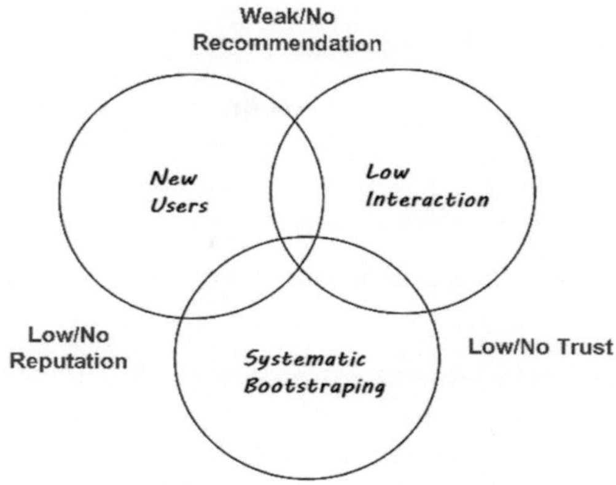

Fig. 6.1 Three reasons for the cold start problem

information based on which the system can depend upon. In this case, the disadvantages of both the new user and new item are present, because of which the techniques used to deal with the two cases are not applicable to system bootstrapping.

In the new item case, whenever a new item is added to a system or catalogue, although it might have content information, it will not have any interactions. So this is the item cold start problem. The problem is mainly created for collaborative filtering algorithms, because the algorithm uses item interactions to make recommendations. So in the condition that there are no interactions, then a purely collaborative algorithm will not be able to recommend an item.

A number of solutions have been devised to tackle the cold start problem. Here in the following sections, we discuss three new types of recommendation systems, namely knowledge-based, ensemble-based, and hybrid systems.

6.3 Knowledge-Based Recommender Systems

Since content-based and collaborative systems both need a lot of information about previous buys and user ratings, hence they are unable to make good recommendations when there is a lack of available data. This is also referred to as the cold start problem.

So they are not suitable for applications in areas where there are highly customized products, like real estate, cars, or other luxury goods. These items are generally bought quite rarely and therefore there are insufficient ratings available in many cases. For example, a person is searching a house with a specific number of rooms, facing a particular direction or in a particular area. So it is unlikely to find a sufficient number of past ratings for a such specific combination of parameters. Similarly, ratings and preferences of cars based on build, engine, color, etc. evolve over time and past ratings may not be suitable for present scenarios as constant improvements are going on in the automobile industry. In such circumstances, knowledge-based recommender systems work very well for products that are not bought regularly, because they rely on explicit

Approach	Conceptual Goal	Input
Collaborative	Give me recommendations based on a collaborative approach that leverages the ratings and actions of my peers/myself.	User ratings + community ratings
Content-based	Give me recommendations based on the content (attributes) I have favored in my past ratings and actions.	User ratings + item attributes
Knowledge-based	Give me recommendations based on my explicit specification of the kind of content (attributes) I want.	User specification + item attributes + domain knowledge

Fig. 6.2 Differences between content-based, collaborative, and knowledge-based systems

user solicitations for their system. The difference in concept between collaborative filtering, content-based, and knowledge-based systems is shown in Fig. 6.2 (Boldi et al. 2008; Bridge et al. 2005).

Knowledge-based recommender systems (Burke 2000; Burke et al. 1996; Felfernig et al. 2007) work well in the following types of scenarios:

- In situations where a user wishes to explicitly specify his/her requirements, interaction is a very important thing in such systems. Such type of detailed feedback is not allowed in collaborative or content-based systems.
- In cases where obtaining the ratings for a particular type of product is difficult, because the product domain is more complex based on the item type and option availability.
- In cases where the ratings maybe time—sensitive, like the ratings on older models of cars or computers or mobile phones, etc. will not be useful beyond a particular time limit because more advanced versions have been rolled out in the market.

So in knowledge-based systems, the user has greater control over the guidance of a recommendation system, because here the user is able to give the specific details for a complex problem domain. Knowledge-based systems can be divided on the basis of their methods of interactions with the users and the respective knowledge bases needed for the interactions. They are constraint-based and case-based recommender systems.

- *Constraint-based recommender systems*—In this type of system, the user normally gives some constraints, like an upper or lower limit on the attributes of the item. Some domain-specific rules are implemented for matching the requirements of the user with the attributes of the user, e.g., a user may search for a car with cruise control and diesel and manual. So the user attributes can also be used in the searching process. Now based on how many results the search returns, the user may modify or relax the constraints if too few results are returned. The above process can be repeated a number of times until the user gets the desired results.
- *Case-based recommender systems*—In this system the user specifies specific cases to be used as target or anchoring points. Then the item attributes that have been defined are used to retrieve similar items based on similarity metrics, which are defined specifically for a domain. So it is the similarity metrics that form the basis of the domain knowledge that is used for the recommendations here. Sometimes the results that are returned by the users are interactively modified to be used as

new targets. Suppose a user gets a result that is almost similar to what the user is searching for, then the user may retry the query but with some modifications in the attributes. Sometimes a directional critique is also used. It is a method where some items with some specific attributes which are less than or greater than the attributes of a particular item are pruned off to help the user through the final recommendations.

The interactive processes of both cases are shown in Fig. 6.3a, b.

The interactions that the users have with the recommender system can be either conversational system, search-based, or navigation-based system. They are explained further as follows:

- *Conversational system*—Here the preferences of the user are found through a feedback loop. This system is useful because when dealing with a complex item domain, the preferences of the user can only be found through an iteration of conversations.
- *Search-based system*—In this system search engines are used to find out user preferences by asking a preset sequence of queries.
- *Navigation-based system*—Here when the user gets a recommended item, he/she will specify the number of changes to be made to it and after an iteration of such change, requests can finally arrive at the desired item. Such systems are also called critiquing recommender systems.

Given the overview, a more detailed discussion of the constraint-based and case-based systems is given in the next sections.

6.3.1 Constraint-Based Recommender Systems

This system allows the users to specify hard requirements or constraints on the item attributes. On addition there will be a set of rules for matching the customer requirements with the item attributes. But it is not necessary that the customers always specify their queries in terms of the same attributes that describe the items. So there also needs to be an additional set of rules that will relate the customer requirements with the item attributes. For example if we take Table 6.1, the following are the customer-specified attributes:

Marital-Status (categorical), Family-size (numerical), Suburban-or-city (binary), Min-Bedrooms (numerical), Max-Bedrooms (numerical), Max-Price (numerical)

These attributes can be either inherent customer properties or they may be customer requirements for the product. These requirements are often specified interactively during a conversation between a customer and a recommendation system. Some of these attributes are also not included in Table 6.1. But while some of the customer requirements like max price may be mapped easily, other mappings like suburban or rural may not be as obvious.

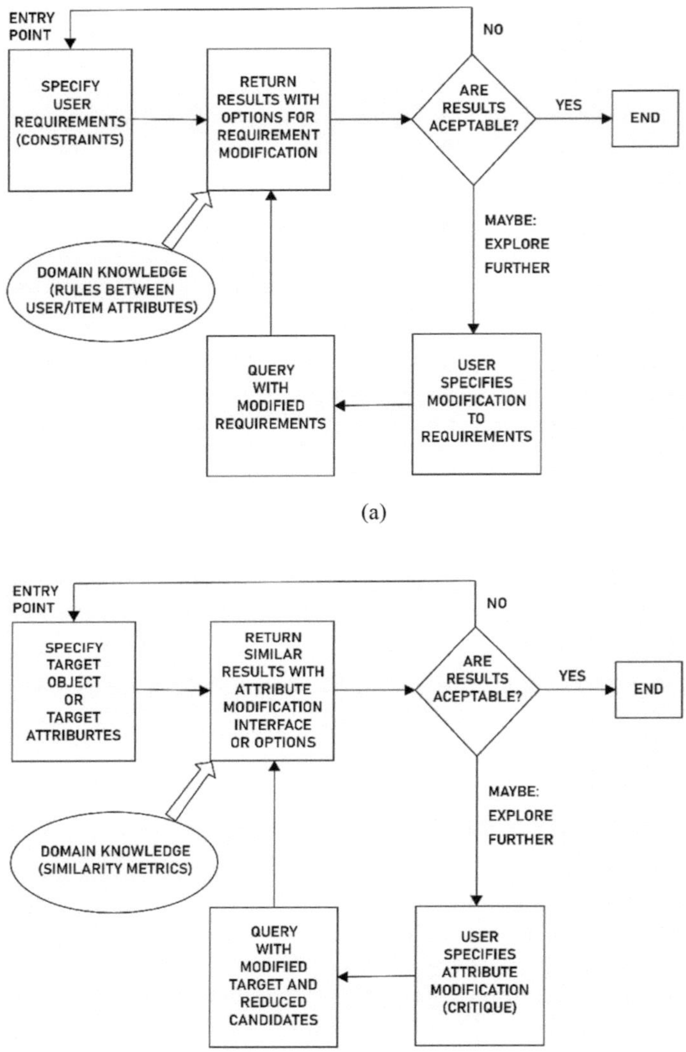

(a)

(b)

Fig. 6.3 **a** Constraint-based interaction, **b** case-based interaction

In a similar way, in a financial application, if a customer specifies a product requirement as *"conservative investments"*, it may be needed to be mapped to some concrete product attributes like *"asset-type = treasuries"*. So there essentially needs to be a way in which these customer attributes/requirements must be mapped into the product attributes, for filtering products for recommendation.

Table 6.1 Examples of attributes for a recommendation app for buying houses

Item id	Beds	Baths	Locality	Type	Floor area	Price
1	3	2	Pune	Townhouse	1600	63 L
2	5	2.5	Chennai	Split level	3600	90 L
3	4	2	Delhi	Ranch	2600	75 L
4	2	1.5	Bangalore	Condo	1500	60 L
5	4	2	Kolkata	Colonial	2700	80 L

This is done by using something called knowledge bases. They contain additional rules that help in the mapping of customer requirements/attributes with product attributes.

Suburban-or-rural = *Suburban* $\Rightarrow$ *Locality* = *<List of relevant localities>*

These rules are called filter conditions, because they map the requirements of the user to the item attributes and then this mapping is used for filtering the retrieved results.

Some compatibility constraints relate customer attributes to one another and are useful when customers give their personal information during an interaction. One such example is as follows:

Marital-status = *single* $\Rightarrow$ *Min-Bedrooms* ≤ 5

So it has been inferred either through data mining of historical data or through domain specific experience that if individuals are single then they do not prefer buying large houses. In the same way, large families will not prefer small houses.

So this constraint can be modeled with the following rule:

Family-Size $\geq 5 \Rightarrow$ *Min-Bedrooms* ≥ 3

6.3.2 Case-Based Recommender Systems

Similarity metrics are used here for retrieving examples that are similar to the cases that are specified. For example in Table 6.1, a user can specify a locality, no. of bedrooms and a preferred location for specifying the set of attributes. But here, there are no hard constraints like minimum or maximum values enforced on the constraints, unlike the constraint-based systems. A similarity function is used for the retrieval of cases which are most similar to the cases specified by the users. So if there are no matches for homes that match the user specifications, then a similarity function is used for retrieving and ranking items that are as similar as possible to the user queries. So in these types of systems do not face the problem of retrieving empty sets. Many differences are also there between constraint-based and case-based recommender systems, in the aspect of how the results are refined. While the former uses requirement relaxation, modification and tightening for the refinement of the results, the latter uses repeated modification of the requirements of the user queries, till a suitable solution was available. This led to the method of critiquing. The basic principle of the critiquing method was that a user could select one or more of the retrieved results and then specify further queries like:

Give me more items like X, but they are different in attribute(s) Y according to guidance Z.

The basic aim of critiquing is to support interactive browsing of the item space so that a user slowly gets aware of further options available to them from the examples that have been retrieved. The advantage of interactive browsing of item space is that a user can gradually learn during the process of formulation of interactive queries. Often in many cases a user may be able to arrive at choices through repeated and interactive exploration, which could otherwise have not been reached at the beginning of the search. If we consider the example of Table 6.1, a user can specify a preferred price, the number of bedrooms and a preferred locality. But it is also possible that a user enters a target address for asking for examples of possible options in houses he/she may be interested in.

By repeating the process of critiquing, the results that a user finally gets are sometimes very different from what the user query had initially specified. This is because of the fact that quite often a user may not be able to easily articulate ALL of the preferred features in the beginning itself. For example, a user may be unaware of the prices for a house with a desired set of features when he/she is first starting the query process. So with the help of this interaction process the gap between the item availability, and the user perceptions is gradually narrowed down.

So for the efficient working of a case-based recommender system, there are two main points that need to be considered while designing the system:

- Similarity metrics, where the importance of various attributes need to be incorporated into the similarity function for the effective working of the system.
- Different types of Critiquing methods are used to provide support to the various goals of exploration.

Similarity Metrics

Suppose there is an application which has d attributes. We need to find the similarity values between two partial attribute vectors defined on a subset S of d attributes (i.e., $|S| = s \leq d$).

Suppose $X = (\times 1 ... \times d)$ and $T = (t1 ... td)$ are two d-dimensional vectors, which are partially specified here and T is the target. We assume that the attribute subset $S \subseteq \{1...d\}$ is specified in both vectors. Partial attribute vectors are used here the queries are usually defined only on a small subset of the attributes that are specified by the user. Then the similarity function $f(T,X)$ between the two sets of vectors is:

$$f(\overline{T}, \overline{X}) = \frac{\sum_{i \in S} w_i \cdot Sim(t_i, x_i)}{\sum_{i \in S} w_i}$$

where *Sim*(*ti, xi*) is the similarity between the values *xi* and *yi,* and weight *wi* is the weight of the *i*th attribute, which regulates the relative importance of that attribute.

Critiquing Methods

The basic idea behind the use of critiques is that in a lot of cases, the users do not often know exactly how to state their query initially. If it is a complex domain, it may become even more difficult for them to translate their requirements in a semantically meaningful way so that they can match with the attribute values of the products. So after seeing the results of a query a user may understand how to phrase her query in a different manner.

Once the users have received the results, there is a feedback mechanism using critiques. Although most cases have interfaces which critique the most similar matching item, a user can also critique any item from the list of retrieved items. Here the users change requests on one or more of the attributes of an item which they like. So in the context of the house buying example in Table 6.1, a user may be interested in a specific house but might want it in a different area or with a different number of bedrooms. So the user can make changes in the features in any one of the products he/she likes. It may be a directional critique (like " less expensive") or a replacement critique (like "different color"). Then the examples which do not meet the user -specified critiques are eliminated and a different set of more similar items are retrieved. If there are multiple critiques, then recent ones are given the higher precedence. So critiques can be of three types, simple, compound, and dynamic critiques. In a simple critique, only a single change to one of the features of a recommended item is done by the user. In a compound critique, a user can specify modification of multiple features in a single cycle. In dynamic critiquing, data mining is used on the results that have been retrieved for finding the most effective roads for exploring and presenting to the user. So dynamic critiques are basically compound critiques as they almost always give combinations of the changes which are presented to the user, with the major difference being that only a subset of the most relevant items is given, on the basis of the recent retrieved results.

6.4 Ensemble-Based and Hybrid Recommender Systems

As already explained in the earlier chapters, different systems have different sources of data and different advantages and disadvantages. While knowledge-based systems need explicit user specifications, content-based and collaborative filtering are based on past ratings and preferences. So knowledge-based systems address the cold start problem in a much better way. But all the models are restrictive when there are multiple sources of data available. If different types of recommender systems are used along with different data sources, then the predictions may be more robust.

This led to the design of Hybrid recommender systems. There are primarily three ways to create such hybrid systems:

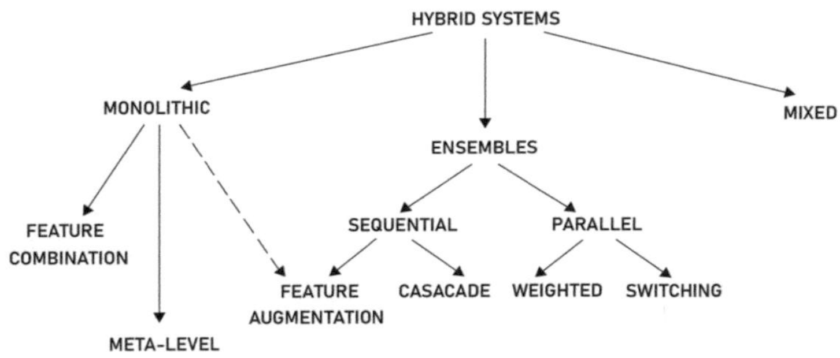

Fig. 6.4 Types of hybrid systems

- *Ensemble Design*—Here, the results from various algorithms are combined into a single and more robust output. So maybe the ratings of content-based and collaborative might be combined together into a single output. But there are differences in the ways the methods are combined together (Ma et al. 2009; Yu et al. 2003).
- *Monolithic Design*—In this system, an integrated recommendation engine is created using various data types. In some cases, the existing CF or CBF methods have to be modified to be fitted to the overall approach, although the two methods are different from each other. It also integrates the data sources very tightly.
- Mixed System—Like the ensemble method, these systems use multiple recommender algorithms, but the difference is that the products that are suggested by the different systems are presented all together and beside each other. For example, the list of programs on television for an entire day is seen as a whole with multiple suggestions. So it's basically the combination of the items that create the recommendation.

So hybrid systems are used in a broader context and while all ensemble systems are hybrid systems, the reverse may not always be true. Figure 6.4 shows the taxonomy of the hybrid systems.

Hybrid recommender systems (Tang et al. 2003; Tran and Cohen 2000; Satten 2005) can be classified as follows:

- *Weighted*—In the weighted system, the combination of the scores from various recommender systems are combined to form a single composite score by using a weighted aggregate of the scores of the separate components of the ensemble. The way for deciding the weights of the components can either be heuristics, or formal statistical models.
- *Switching*—In this method the algorithm switches among the different types of recommender systems, depending on the requirements of the system then, e.g., a knowledge-based recommender system may be used in the early phases to combat the challenge of the cold start problem. Then gradually as the system gathers more ratings, the system might switch to another system like CF or CB algorithms, or whichever algorithm will be more suited at that point in time.

- *Cascade*—Here the suggestions given by one recommender system may be refined by another recommender system. So the training process from one recommender system is biased by the output it receives from the previous system. The final result is shown as one single output.
- *Feature augmentation*—When the output from one recommender system is fed to another type of recommender system, the cascade technique refines the recommendations provided by the previous system, after which the feature augmentation technique is used to make those features as input for the next system. This method is somewhat similar to stacking, a method used often in classification. This approach is more of an ensemble method than a monolithic method.
- *Feature Combination*—Here a combination of the features from the various sources is used for a single recommender system. This is a monolithic system, and not ensemble based.
- *Meta level*—When the output from one recommender system is used as the output for another recommender system, the usual order of the choice of the methods is that of content-based followed by the collaborative system. So the content features of the collaborative system are used to find out similar user groups. After that, the rating matrix is used in conjunction with similar user groups to arrive at the recommendations. So the collaborative system needs to be modified here so that a content matrix can be implemented for a finding of peer groups. So using the meta-level approach is more of a monolithic approach rather than an ensemble system. They are also sometimes called "collaboration via content" based on the order in which the methods are used.
- *Mixed*—Here the recommendations from various engines are suggested to the user together but not as a combination. When the result is a composite entity, then it is sometimes needed to recommend multiple items as a related set. So mixed systems neither fall in the ensemble nor monolithic category. They are in a different category by themselves.

So the first four are ensemble-based categories and the next two are monolithic, and the last one is a mixed system.

6.5 Ensemble Methods from the Classification Perspective

Ensemble methods are applied to the area of data classification for increasing the robustness of the learning algorithms, but it is also applied to other types of recommender systems also. Collaborative filtering and classification methods differ from each other in the aspect that the class variables are not clearly defined in the first method and there may be missing entries in any column or row. The missing rows also indicate that training and test instances are not clearly defined. Now, one may ask if the bias-variance theory for classification is also applicable for recommender systems. It has been observed that the combination of different collaborative recommender systems give a higher degree of accuracy in results. The reason for this is that the bias-variance theory, which was designed for classification, is also applicable for collaborative filtering areas. Therefore most of the traditional ensemble techniques from classification may be generalized to collaborative filtering. The only problem that arises is that if the missing entries occur in row or column of data, it may be algorithmically challenging to generalize the ensemble algorithm for classification to collaborative filtering. Let us consider a classification or regression model where we need to predict a specific field. Then the classifier error for the prediction of the dependent variable can be broken into three parts, namely the bias, variance, and the noise:

Bias—All classifiers make their own assumptions for modelling regarding the type of the decision boundary between the classes. In case a classifier has a high bias, consistently incorrect predictions will be made by it on specific choices of test instances around the incorrectly modeled decision boundary, and this holds true even if the training data samples are different during the learning process.

Variance—If there are random variations in the selection of the training data, then it will lead to dissimilar models. In that case there may be inconsistencies in the prediction of the dependent variable for a test case, for different selections of training data sets. The variance of a model is also closely linked to overfitting. If a classifier has a tendency to overfit, then the predictions it makes will also be inconsistent for the same test case but with different sets of training data.

Noise—The intrinsic errors in labeling the target class is called noise. However not much can be done for correcting it, because noise is an intrinsic property of the quality of data. So ensemble analysis normally focuses on the reduction of variance and bias.

The formula for the expected MSE of a classifier is:

$$Error = Bias^2 + Variance + Noise$$

The total error of a classifier can be reduced by reducing either the bias or the variance. The classification and the collaborative filtering methods differ in the fact that the missing entries may be present in any column as against only in a class variable. But the result of the bias-variance is valid even when it is applied for the prediction of a particular column, irrespective of whether they are specified incompletely or not. So the rules of ensemble analysis hold good for collaborative filtering.

6.6 Summary

In this chapter, we have introduced three new types of recommendation systems, namely knowledge-based, ensemble-based, and hybrid systems. The algorithms in the previous chapters used past ratings of users for coming up with suggestions. But in the special cases where sufficient amount of past information is not available, or where there are complex variations in the combinations of preferences, or there are some less frequently, highly customized or high value items to be bought, the methods in the previous chapters were not very helpful. We have analyzed the types of applications that need user interactions for the recommender systems to arrive at their suggestions instead of using previous user ratings or finding users with similar preferences. An overview of knowledge-based recommender systems and their ideal application areas are discussed along with the various types of hybrid and ensemble-based systems.

Since recommender systems rely on huge amounts of data to be able to make more accurate predictions, a successful system should be able to efficiently handle huge volumes and varieties of data, or what we now call Big Data. The next chapter gives an overview of the big data behind such recommendation systems, its roles, and challenges.

Think Tank

1. *What are knowledge based recommender systems?*
2. *What are ensemble based systems?*
3. *What are the different types of hybrid recommender systems?*
4. *What is the relation between variance, bias and noise?*

References

Aggarwal C, Procopiuc C, Yu PS (2001) Finding localized associations in market basket data. IEEE Trans Knowl Data Eng 14(1):51–62

Bobadilla J, Ortega F, Hernando A, Gutierrez A (2013) Recommender systems survey. Knowl-Based Syst 46:109–132

Bohnert F, Zukerman I, Berkovsky S, Baldwin T, Sonenberg L (2008) Using interest and transition models to predict visitor locations in museums. AI Commun 2(2):195–202

Boldi P, Bonchi F, Castillo C, Donato D, Gionis A, Vigna S (2008) The queryflow graph: model and applications. In: ACM Conference on Information and Knowledge Management, pp 609–618

Bridge D, Goker M, McGinty L, Smyth B (2005) Case-based recommender systems. Knowl Eng Rev 20(3):315–320

Burke R (2000) Knowledge-based recommender systems. In: Encyclopedia of library and information systems, pp 175–186

Burke R, Hammond K, Young B (1996) Knowledge-based navigation of complex information spaces. In: National Conference on Artificial Intelligence, pp 462–468

Felfernig A, Teppan E, Gula B (2007) Knowledge-based recommender technologies for marketing and sales. Int J Pattern Recogn Artif Intell 21(02):333–354

Ma H, Lyu M, King I (2009) Learning to recommend with social trust ensemble. In: ACM SIGIR Conference, pp 203–210

Tang T, Winoto P, Chan KCC (2003) On the temporal analysis for improved hybrid recommendations. In: International Conference on Web Intelligence, pp 214–220

Tran T, Cohen R (2000) Hybrid recommender systems for electronic commerce. In: Knowledge-Based Electronic Markets, Papers from the AAAI Workshop, Technical Report WS-00-04, pp 73–83

van Satten M (2005) Supporting people in finding information: hybrid recommender systems and goal-based structuring. Ph.D. Thesis, Telemetica Instituut, University of Twente

Yu K, Shcwaighofer A, Tresp V, Ma W-Y, Zhang H (2003) Collaborative ensemble learning. Combining collaborative and content-based filtering via hierarchical Bayes. In: Conference on Uncertainty in Artificial Intelligence, pp 616–623

Chapter 7
Big Data Behind Recommender Systems

Abstract For recommender systems to make accurate predictions about the preferences of users, they rely on varieties of information and feedback from the customers. So this naturally involves dealing with and processing huge volumes of data every day. So here we see the concept of big data and why it is so important. We also see how recommender systems can benefit from using big data, what the types of data stored and what the challenges are. Finally, some examples show how exactly it is used by the recommender systems by taking the example of Twitter.

Keywords Big data · Unstructured data · Semi-structured data · Pre-processing · Variety · Volume · Velocity of data · Singular Value Decomposition (SVD)

7.1 Introduction

Nowadays a majority of companies use their own versions of recommendation systems to provide more accurate and customized recommendations to their users. A recommendation engine applies its algorithms to different types of customer data to arrive at personalized recommendations and suggestions. For this, a recommendation system needs a huge amount of data to be able to perform its job effectively. So it relies heavily on the supply of plenty of user data on big data. The user data can be past purchases, browsing history, as well as feedback that is provided to the recommendation systems so that they can provide an effective and relevant recommendation to the users. In this chapter, we take a look at what is big data, what are the types of data that are used by recommendation systems, how big data is used in recommendation systems, and the existing issues/challenges that the implementing agencies face while using these data for their applications. We also consider the example of the role of big data in Netflix, to show how important a role big data plays in building successful recommendation systems.

P. Kar et al., *Recommender Systems: Algorithms and their Applications*, Transactions on Computer Systems and Networks, https://doi.org/10.1007/978-981-97-0538-2_7

7.2 What is Big Data?

Although the concept of big data is relatively new, large data sets originated way back in the 1960s and 70s, with the setting up of data centers and relational databases. It was around 2005 that people started to realize the huge volumes of data that were being generated through Facebook, YouTube, and other online services. It was around this time that Hadoop (Dobrea and Xhafa 2014; Sheikh 2013) was developed which was an open source framework for storing and analyzing big data sets. Another similar framework was NoSQL. With the advent of such open source frameworks, it became easier to store and work with big data (Philip Chen and Zhang 2014; Chardonnens 2013). But it is not only humans who are generating such huge amounts of data. After the rise of the Internet of Things (IoT), there were many objects and devices which were connected to the Internet which were gathering data on customer usage patterns and product performances. The introduction of machine learning has brought in even more volumes of data.

Big data is defined as data that contains greater variety, arriving in increasing volumes and with more velocity. These are also known as the three Vs in big data. Big data consists of larger and more complex data sets, especially from new data sources. Since these data sets are so voluminous, therefore the traditional data processing software can't manage them. But if these massive volumes of data can be harnessed in some way, they could be hugely successful in addressing business problems that we could not handle before.

Big data is characterized by the 3Vs, volume, velocity, and variety, as mentioned in Fig. 7.1. The first parameter Volume implies that there are huge volumes of data involved. Most of the data is low-density and unstructured data. A lot of this data is also of unknown value and for many organizations can run into hundreds of petabytes of data.

The velocity of the data means the rate at which this data is received and in a lot of cases how quickly this data can be processed and acted upon. Usually the highest velocity of data streams directly into the memory instead of being written to the disk.

Fig. 7.1 Five Vs of big data

Some Internet-enabled smart products operate in real-time and need an evaluation of data on a real-time basis.

Variety refers to the numerous types of data that are available and need to be processed and most of it is unstructured. The traditional data used to be structured and could be fitted into a relational database. But with the advent of big data, this was no longer possible as data now comes in new unstructured data types. The various types of unstructured and semi-structured data types like text, audio, and video need additional pre-processing before some meaning can be derived from it and metadata can be supported.

Recently two more Vs have emerged, value and veracity. The data may have intrinsic value but it is actually of no use unless that value is discovered. Similarly, it is important to know how trustworthy the collected data is, which is called veracity.

Big data works in three steps: integrate—it brings together data from various disparate sources and applications, which need new techniques as the traditional systems are unable to process them, manage—big data requires storage solutions, like on the cloud or on-premises, and analyze—building data models and applying ML and AI for analyzing the collected data. Big data has a wide array of application areas ranging from product development, predictive maintenance, customer experience, fraud and compliance, machine learning, operational efficiency, and drive innovation.

In the next section, we see the role big data plays in recommender systems.

7.3 How Big Data is Used in Recommender Systems

Recommender systems are one of the most common and easily understandable applications of big data. They are based on well-defined and logical phases of data collection, ratings, and filtering. We can find applications of recommendation systems in eCommerce, entertainment, gaming, education, advertising, home decor, and some other industries. The applications differ based on the types of recommendation services they provide but the central goal of all of them is to personalize content and offers.

In order to achieve this, machine learning (ML) engineers have designed recommender systems that redefine the ways customers search for products or services or learn about new opportunities and goods they may be interested in. ***However, the driving force behind all these systems is big data.*** There are at present numerous types of recommender systems designed to offer a variety of personalized content, but the work of all of them is based on voluminous datasets.

As already mentioned in Chap. 3, there are three major types of recommender systems:

- Content-based filtering
- Collaborative filtering
- Hybrid recommender systems

All of these are reliant on user behavior data, including activities, preferences, and likes, or can take into account the description of the items that users prefer, or both. It has been widely observed that the incorporation of recommender systems into businesses has increased the number of items sold, has sold more diverse items, is better at increasing user satisfaction, and helps the service providers to have a better understanding of what the user wants. It also helps the user to make customized and informed decisions thereby increasing customer loyalty and retention. However, even the most advanced of these recommender systems can be rendered irrelevant and ineffective without the presence of big data. A recommendation system can't do its work if it is not supplied with sufficient data for the algorithms it uses, because such systems rely heavily on information about past purchases, browsing history, and feedback from a huge number of customers. Such huge volumes of data can only be provided with the help of big data.

In the next section, we discuss the diverse types of data that are required by a majority of recommendation systems.

7.4 Types of Data Used in Recommender Systems

Recommender systems base their algorithms on a variety of user data in order to arrive at customized recommendations. Some examples are personalized product recommendations, website personalization, real-time notifications, and personalized loyalty programs and offers. For recommending personalized products, the engines assist in understanding the preferences of each customer who visits the site and shows the most relevant types of products to the customers. Website personalization helps to increase sales by dividing the customers into segments and targeting them with relevant real-time messages and offers. By providing real-time notifications, recommendation systems instill a sense of trust in the customers by making the presence of the company felt. It also helps in pushing personalized loyalty programs and offers to its diversity of customers.

In order to do all these, a recommendation system must search the web to collect a variety of data. They can be broadly categorized as:

- *User Behavior data* (*historical data*): This type of data is collected by monitoring the logs of on-site activities like clicks, searches, pages visited, and item views. It also needs to track off-site activities like clicks in emails, mobile applications, and in their push notifications.
- *Particular Item details*: The recommendation engines also need to collect the details of information about the particular items used by a customer, like an item name, the category of the item, its price, and description.

- *Contextual Information*: The engines also need to gather contextual information like the devices used by the customers, their current locations, and their referral URLs.

In the next section, we take a look at the types of challenges faced by the service providers while implementing big data in their recommendation systems.

7.5 Challenges

There are however several types of challenges (Río and López 2014; Alejandro Zarate Santovena 2013; Vemuganti 2013) that are encountered by organizations in the usage of this big data for their advantage. One of the primary reasons was that of protecting the user's privacy. A lot of privacy issues were raised when companies tried to collect and process a user's personal data because they claimed that the identity of the user could be traced from the data collected. Some other major challenges are:

Data Capture and Storage—The size of data sets is growing size everyday, because the sources of the data are now manifold, e.g., through mobile devices, sensors, remote sensing, software logs, cameras, microphones, RFIDs, and many more. Around 2.5 quintillion bytes of data are created everyday, and this is increasing exponentially everyday. Moreover, the data is mostly unstructured data from a variety of sources. The collection of all this data is an expensive process, but in many cases, the data has to be just deleted, mainly because of insufficient space to store them. To store and analyze the data properly, new techniques and frameworks have to be designed, as the existing relational databases, etc. are not adequate to handle them anymore.

Data Transmission—Cloud data storage is the most popular way of storing huge volumes of data. However network bandwidth capacity is a major bottleneck when it comes to accessing this data from the cloud, especially when the volume of data is large. Moreover, storing the data in the clouds also poses several security risks which also need to be dealt with.

Data Curation—Data curation is aimed at data discovery and retrieval, data quality assurance, value addition, reuse, and preservation over time. It involves authentication, archiving, management, preservation, retrieval, and representation. However, the existing database management tools are inadequate for processing this big data. In addition, the volume of the data will only increase in the future, as more and more organizations are realizing the benefits of big data in analyzing business trends, preventing diseases, and combatting crime. So newer technologies are being used to tackle this.

Data Analysis—In some applications like navigation, social networks, finance, biomedicine, and intelligent transport systems, the time to analyze the data should be as less as possible because they need the results almost on a real-time basis. But when dealing with huge volumes of data, reducing the latency in the analysis is a big challenge.

Data Visualization—The primary aim of data visualization is to represent knowledge more intuitively and effectively through the use of various types of graphs. Presenting the data in a schematic form is much more intuitive and easier to comprehend. However, for big data applications, it is difficult to perform data visualization, because of its large size and high dimension. The existing big data visualization tools perform poorly in terms of functionalities, scalability, and response time.

Sparsity—Data sparsity is a major challenge to the recommender system. In such a situation, the number of items to be rated are much more than the already rated items by the user. Thus, the very few entries of the user-item matrix are filled with values that results the matrix to be sparse and reduced recommendation. One possible solution to this problem is to provide suggestion to a user by checking his profile and analyzing similarities so that if two users share a common interest on a product then one user's recommendation can be given to another user. The sparsity problem is addressed by Singular Value Decomposition (SVD) technique, which reduces the dimensionality of sparse rating matrix.

Scalability—With the increase in no of users, products, and rating the scalability issue arises in recommender systems. When the product information and number of items increase as well as recommender systems are expected to quickly generate recommendations to the customers the system requires an increased scalability. But the execution of such systems becomes strenuous and exorbitant. So it is essential to design an efficient and effective data model that can adapt itself to deal with growing dataset. One possible solution to the scalability issue is to perform computation on multiple machines in parallel using distributed algorithm.

Overspecialization—Due to overspecialization issue of recommender systems, a highly rated item is suggested which has already purchased or experienced by the user. As it doesn't work according to the user preference, the user lose interest in the system. Neighborhood collaborative filtering, randomness introduction using genetic algorithm, or by removing similar items are proposed to handle overspecialization.

Serendipity—Each recommender system should achieve the very crucial Serendipity objective that focuses on achieving user trust and loyalty. Recommender systems should provide significantly novel and relevant suggestions in contrast to user's previous rating for items. It is challenging to apprehend the idea of serendipity completely due to its subjective nature and is rarely seen in real life scenarios. Solutions like re-ranking the accuracy results has been introduced to achieve serendipity.

Coverage—As the types of items for cataloging increase, the systems need to have a high coverage but maintain low latency.

Diversity—The recommendation engine should be able to give its users a variety of recommendations.

Adaptability—The system should be able to adapt quickly to the continually changing world of content.

User Preferences—The system should be capable of handling users of varied interests within one ranking framework.

In the next section, we take the example of Twitter to understand how a successful recommendation system can benefit from big data.

7.6 An Example of the Role of Big Data in Twitter

Twitter as everyone knows is a micro-blogging site where hundreds of millions of users post tweets everyday. The most commonly seen aspects on Twitter are "Who to follow" and "Trends for you". The core engine that runs in the background is a recommendation engine that relies on big data to arrive at such suggestions. However, there are some differences between Twitter and non-Twitter recommendations that we discuss here. There are three main recommendation products that Twitter pushes to its customers:

Users to be followed—This aspect suggests the usernames of users who a person may be interested in following, and the total number of such recommendations may well reach 1 billion. These recommendations have a longer shelf life meaning they are valid for longer durations of time.

Tweets—The tweet recommendations that a user gets in his/ her feeds can be of the order of a few hundred million, within a few hours. The shelf life of these tweet suggestions is less these days, because of fast-changing news everyday.

Trending/Events—Trending has the least number of recommendations as compared to the above two, as well as very short shelf life. This is because a lot of the users may be following similar trends.

In comparison, many non-Twitter recommendation engines like movie suggestions or product suggestions have a longer shelf life. The advantage of having a longer shelf life is that, since we are dealing with higher volumes of data, it takes more time to analyze and process such data. So the latency or turnaround time will be more. But for recommendations with shorter shelf life, the analysis of huge volumes of data needs to be done almost on a real-time basis to remain accurate and relevant.

On Twitter, there are two types of recommendations, user-user, and user-item.

In the user-user recommendations, the similarities are found based on who the users follow and are not very robust because a person may follow another person for a particular ideology, and another person having a common follower might not necessarily have the same ideology.

The user-item recommendation technique is based on the likes /dislikes or interests of a user. While the long-term interests of a user may not vary, the short-term interests may vary and may be more difficult to recommend within such short periods.

Twitter uses both collaborative filtering as well as content-based filtering, and sometimes a hybrid of the two models for their recommendations. The content-based filtering method generates more content for processing (Xhafa and Barolli 2014; Tran 2013).

7.7 Singular Value Decomposition-Based Recommender Systems

Data sparsity is a major issue in recommender system. Here we discuss the solution to deal with the sparsity issue in recommender system with an example.

7.7.1 Singular Value Decomposition (SVD)

SVD is a popular matrix decomposition technique that decomposes a matrix M of $k \times l$ and rank r into three matrices U, S, and V such that,

$$M = U \cdot S \cdot V^{T} \tag{7.1}$$

where

U is an orthogonal matrix of size $k \times r$ that holds left singular vectors of M in its column. This means r columns of U hold eigenvectors of the r nonzero eigenvalues of MM^{T}.

S is a diagonal matrix of size $r \times r$ that holds the singular values of M in its diagonal entries in decreasing order such as $s_1 \geq s_2 \geq s_3 \geq \ldots\ldots \geq s_r$, which are nonnegative square roots of the eigenvalues of MM^{T}.

V is an orthogonal matrix of size $l \times r$ that holds the right singular vectors of M in its columns, which means its r columns hold eigenvectors of the r nonzero eigenvalues of $M^{T}M$.

Additionally, S could be reduced by taking the largest n singular values only and thus obtain S_n of size $n \times n$. Similarly, U and V could be replaced by keeping the first k singular vectors and discarding the rest resulting U_n of size $k \times n$ and V_n of size $k \times n$. As a result, $M_n = U_n \cdot S_n \cdot V_n^{T}$ and $M_n \approx M$, where M_n is the closest rank n approximation to M.

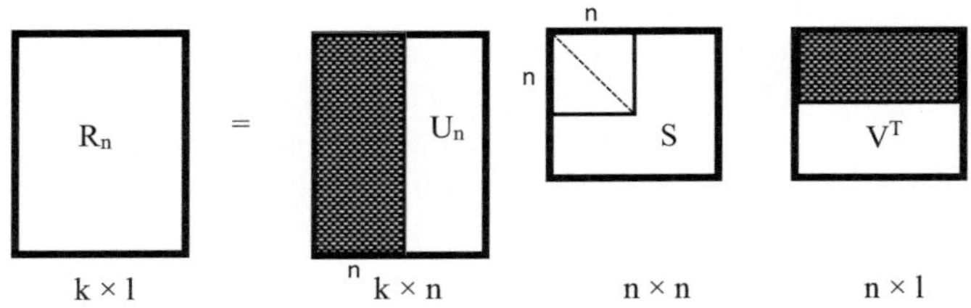

7.7.2 Recommender Systems Using SVD

Relationship between users and items and similarity between them could be induces by some lower dimensional structure in the data to the recommender system using SVD. As an example, the ratings provided by a certain user to a particular item such as movie depends on some implicit factors like the preference of that user across different movie genres (Reeve 2013; Popescu and Etzioni 2007). The SVD-based recommender system considers users and items as unknown feature vectors to be learnt by applying SVD to user–item matrix and breaking it down into three smaller matrices: U, V, and S. This is done by constructing the sparse user-item matrix from the input data set and then imputing it by some values to fill the missing ratings and reduce its sparseness before computing its SVD. There are several imputation techniques: impute by Zero, impute each column by its Item Average, impute each row by its User Average, or impute each missing cell by the mean value of User Average and item average. This is resulted in a filled matrix M_{fl} that could be normalized by subtracting the average rating of each user from its corresponding row resulting in M_{nr}, which is useful in offsetting the difference in rating scale between the different users. Here, SVD could be applied to M_{nr} to compute U_{n} (this holds users' features), S_{n} (holds the strength of the hidden features) and V_n (holds items' features) such that their inner product will give the closest rank-k approximation to M_{nr}. The SVD removes noise from the user-item relationship by discarding the small singular values from S through lower-rank approximation of the user-item matrix that is better than the original technique. Hence, the dot product of the corresponding feature vectors are computed to predict the reference of user i to item j. This means, compute the dot product of the ith row of $(U_n{\cdot}S_n)$ and jth column of V_k^T and add back the user average rating that was subtracted while normalizing M_{fl}. This is presented as:

$$P_{ij} = \overline{r_a} + (U_n \cdot S_n)_{i,\_} \cdot V^{\mathrm{T}}_{\_,j} \tag{7.2}$$

where P_{ij} is the predicted rating for user i and item j, $\overline{r_a}$ is the user average rating, $V^T_{\_,j}$ is the jth column of $\mathrm{V^T}$ and $(U_n.S_n)_{i,\_}$ is the ith row of the matrix resulting from multiplying U_n and S_n. In point of fact, the dot product of two vectors measures the cosine similarity between them. Thus, the above formula could be interpreted as finding the similarity between user i and item j vectors and then adding the user average rating to predict the missing rating P_{ij}. The Algorithm 7.1 presents the technique of movie recommendation by a large-scale SVD-based Recommender System.

Algorithm 7.1 Large Scale SVD-based Recommender System for Movie Recommendation

Input: MovieLense Big Data Set (B), dimensions (n)
Output: filled user-item matrix P with predictions to all the originally missing ratings read in the form of tuples (userID, itemID, rating)

```
for ∀userID ϵ TrainingSet do ▷ Construct user-item matrix
    r[] ← all the rating given by the user with userID
    M[][] ← r[]
end for
for ∀row(r) ϵ M do                                  ▷ Impute M
    v1 = average (r[])
end for
for ∀column(c) ϵ M do
    v2 = average (c[])
end for
for ∀Mij ϵ M do
    if Mij == NULL then
        Mij ← average(v1, v2)
    end if
end for
for ∀row(r) ϵ M do                               ▷ Normalize M
    for ∀Mj ϵ r do
        Mj ← Mj − v1
    end for
end for
U, S, V ← computeSVD(R, n)
US ← dotProduct(U, S)          ▷ Predicting missing ratings
V^T ← transpose(V)
P ← dotProduct(US, V^T)
for ∀row(r) ϵ P do                             ▷ De-normalize P
    for ∀Pj ϵ r do
        Pj ← Pj + v1
    end for
end for
return P
```

7.8 Summary

Thus in this chapter, we have seen what is Big data, and what are its features. We have also seen what are the different types of big data and what are the challenges involved. We have also seen how big data plays a very important role in the development of a successful and relevant recommendation system. In the next chapter, we discuss what how to build trust-centric and attack resistant recommendation systems.

Think Tank

1. *What is Big Data?*
2. *What are the five Vs in Big Data?*
3. *What are the different types of data in recommender systems?*
4. *What are the challenges in big data?*

References

Chardonnens T (2013) Big data analytics on high velocity streams: specific use cases with storm. Software Engineering Group, Department of Informatics, University of Fribourg, Switzerland

del Río S, López V, Benítez JM, Herrera F (2014) On the use of MapReduce for imbalanced big data using Random Forest. Research Center on Information and Communications Technology, University of Granada, Granada

Dobrea C, Xhafa F (2014) Intelligent services for Big Data science. Future Generat Comput Syst 37:267–281

Philip Chen CL, Zhang C-Y (2014) Data-intensive applications, challenges, techniques and technologies: a survey on Big Data. Inform Comput Sci Intell Syst Appl 275:314–347

Popescu A-M, Etzioni O (2007) Extracting product features and opinions from reviews. In: Natural Language Processing and Text Mining. Springer, London

Reeve A (2013) Big data integration data integration, best practice techniques and technologies. Morgan Kaufmann, pp 141–156

Santovena AZ (2013) Big data: evolution, components, challenges and opportunities. Ms thesis, Massachusetts Institute of Technology, Sloan School of Management, Cambridge

Sheikh N (2013) Big data, Hadoop, and cloud computing, implementing analytics. Morgan Kaufmann

Tran VT (2013) Scalable data management systems for Big Data. Ph.D. thesis, École Normale Supérieure de Cachan, Antenne de Bretagne, INRIA, Rennes

Vemuganti G (2013) Challenges and opportunities. Infosys Labs Briefings 11(1)

Xhafa F, Barolli L (2014) Semantics, intelligent processing and services for big data. Int J Grid Comput Sci 37:201–202

Chapter 8
Trust-Centric and Attack-Resistant Recommender System

Abstract In the information age where social media and online platforms are popular, recommendation systems are being used more and more widely. As a result, the evaluation of recommendation systems has become more rigorous, and a better-quality recommendation system can largely improve the competitiveness of products. This chapter proposes three different ways to improve the performance of recommendation systems based on both attack and trust: usage of the Simulated Annealing Algorithm in the score propagation model, evaluation and usage of users' activeness, and some administrative measures.

Keywords Attack · Attack modeling · Attack detection · Trust · Trust scoring · Trust-based recommender systems · Score propagation model

8.1 Introduction

The recent several decades have witnessed the prosperous development of information and communication technologies, which contribute to the interconnection of the world. As a consequence, the amount of daily online activities is boosting exponentially and a large volume of data is being generated every second. Meanwhile, heterogeneous online items including commodities, short videos, music, news, etc., are emerging at an unprecedented speed. When all kinds of data about the users and the items are fed into a system, the system will be confronted with the problem of information overload when it is trying to recommend items to the users. To mitigate this problem, many recommender systems are being developed. The two most representative recommender systems are the collaborative filtering recommender system and the content-based filtering recommender system (Sridevi et al. 2016). Besides these, knowledge-based and hybrid recommendation techniques are also frequently applied. However, some recommender systems may be vulnerable to attacks, especially collaborative filtering recommender systems. Therefore, some techniques are required to improve the efficiency of the recommender systems in the presence of attacks.

P. Kar et al., *Recommender Systems: Algorithms and their Applications*, Transactions on Computer Systems and Networks, https://doi.org/10.1007/978-981-97-0538-2_8

Attack and trust are two significant terms in the recommender system. With the advent of the information age, the amount of data is gradually increasing, and people need to spend a lot of time filtering useful information, which is called *Information Overload*. Although collaborative filtering alleviates the information overload phenomenon to a large extent, its own openness makes it vulnerable to attacks leading to inaccurate recommendation results. Therefore, it becomes critical to accurately identify the attacker and determine the level of trust in the target user.

Behind the explosion of the recommendation system is the current situation of information overload, people need to help themselves find the right content through the algorithm. The ability to accurately identify user needs is the core of evaluating a recommendation system. One of the quantitative ways to assess this includes the robustness of the system. Flexibility in adjusting the trust rating of the recommender system network and timely identification of attackers increase robustness.

8.2 Literature Review

8.2.1 Concept of Trust in Recommender Systems

Trust is a complicated and abstract notion that is defined differently in different disciplines. Nonetheless, there has been a fundamental and necessary consensus on trust since the dawn of human society: trust is the foundation for all value exchange. Finding a set of suitable criteria to assess trust in a recommender system is therefore critical for the system's overall success.

8.2.1.1 Definitions of Trust

The definition of trust falls into different categories, and in many cases, its exact definition is quite ambiguous. The quantitative study of trust had proliferated since 1992 when Marsh pioneered a formal mathematical model of trust and introduced the concept of computational trust (Marsh 1994). Golbeck and Hendler (2005) defined trust as: "a commitment to believe in the smooth running of the future actions of another entity" in computer science. In other words, item *A* trusts entity *B* if *B* satisfies the requirements or passes the tasks of *A* (Golbeck 2006).

8.2.1.2 Properties of Trust

In social networks, trust has the following properties (Golbeck and Hendler 2005):

- Asymmetric: if user *A* trusts user *B*, this doesn't mean that *B* trusts *A*.
- Not distributive: if user *A* trusts user *B* and *C*, this doesn't mean that *A* trusts *B* and *A* trusts *C*.

- Not generic: if user A trusts user B in field X, this doesn't mean that A trusts B in field Y.
- Transitivity: we assume that trust can be transitive in the recommender system under certain constraints.

8.2.1.3 Modeling Trust in Recommender System

Trust-based recommender system models can be classified into two categories based on the ability to use the trust context approach. Context refers to the user's domain of expertise and aims to consider trustworthy relationships between users of the user's skills in a given context (Selmi et al. 2016).

- Approach with trust contextualization
  1. Model of Abdul-Rahman

     This model defines trust as a subjective measurement or conviction about an individual's experience in a given context. This belief selects a value among four values based on the user's opinion: very bad, bad, good, and very good. For Abdul-Rahman, the trust between two users is determined only by their interaction, since transferability is not considered (Abdul-Rahman and Hailes 1998). Figure 8.1 shows the corresponding meanings and descriptions for different values defined by the model of Abdul-Rahman.
  2. Model of Charif Alchiekh Haydar

     In this in-context model, User X decides whether or not to collaborate with the target user based on the target user's reputation. The keywords collected from the question to which the target offers the proper response determine the user's reputation score. As a result, the user profile is built on keywords, and the user's reputation score isn't fixed. A node is generated between a user and each term linked with a query when he offers an appropriate answer. The new model aims to order the list of users after establishing the reputation scores of the different users answering a question, and the user with the greatest reputation score will deliver the answer, which is more likely to be reliable under this forecast (Haydar et al. 2013).

Value	Meaning	Descirption
-1	Distrust	Completely untrustworthy.
0	Ignorance	Cannot make trust-related judgement about entity.
1	Minimal	Lowest possible trust.
2	Average	Mean trustworthiness. Most entities have this trust level.
3	Good	More trustworthy than most entities.
4	Complete	Completely trust this entity.

Fig. 8.1 Model of Abdul-Rahman

- An approach without trust contextualization

1. MoleTrust

   This model treats trust as a binary value. Trust propagation is a fundamental property in MoleTrust and the propagation of trust is represented by transmissibility. It is assumed that the degree of trust between *X* and *Z* is not equal to the degree of trust between *X* and *Y*. In fact, when user *X* trusts user *Y*, and user *Y* trusts user *Z*, then *X* trusts *Z*. The maximum trust distance between the two users is 4 defined by the author for fixing the trust propagation (Massa and Bhattacharjee 2004).

   The trust value between to users *X* and *Z* can be predicted by Eq. 8.1:

$$\mathrm{tr}(X, Z) = \begin{cases} \frac{(d-n+1)}{d} & \text{if } n \leq d \\ 0 & \text{if } n > d \end{cases} \tag{8.1}$$

2. TidalTrust

   This model is used in social networks. Its purpose is to recommend movies to users. In this model, a user can evaluate his trust in another user using a discrete value of [1, 10], and, in addition, each user can rate the movie with 1–5 stars. The trust network between users is represented by a directed graph (Golbeck and Hendler 2005).

   The recommendation score r_{sm} calculated by a source s of a movie m is calculated using Eq. 8.2:

$$r_{\mathrm{sm}} = \frac{\sum_{i \in S} t_{\mathrm{si}} r_{\mathrm{im}}}{\sum_{i \in S} t_{\mathrm{si}}} \tag{8.2}$$

3. Model of O'Donovan

   This model is mainly based on collaborative filtering. In this new model, consumers and producers refer to users and neighborhoods. A new layer of trust in collaborative filtering by changing the used keywords is added which is the main principle of this model. There are three methods of adding a trust layer: weighting, filtering, and combination (Haydar 2014).

   The first way involves replacing the similarity in collaborative filtering with the value *w* (*c*, *p*, *i*) where *c* stands for consumer, *p* for a producer, and *i* for an item. It is formalized using Eq. 8.3:

$$w(c, p, i) = \frac{2 \times \mathrm{similarity}(c, p) \times \mathrm{reputation}(p, i)}{\mathrm{similarity}(c, p) + \mathrm{reputation}(p, i)} \tag{8.3}$$

4. Model of Simon

   This model is a social recommendation system based on social relationships between users. In this system, users only communicate with trusted users. Trust is therefore an explicit value. The system aims to predict the missing portions between users and objects, which are called scores (Meyffret et al. 2012a, b).

8.2.2 *Social Scoring*

Trust-based recommender system is applied widely in social networks. For example, social scoring is an essential part of many communication applications. The measured impact of your social media accounts on the accounts of other social media users is known as social media scoring. In order to estimate your social media score, this method considers the following properties:

- Friendships or followers count.
- Per day, week, month, or year, the amount of content that is published.
- The number of people who remark on your postings on social media.
- The number of times your social media account interacts with the accounts of others.

In a social network, the term friend refers to an actor's direct relationship. A social relationship is an undirected data-sharing connection between two acquaintances in a social network. The worth of information offered depends on the amount of trust between two friends, which goes from 0 to 1. The higher the level of trust, the more valuable the information provided.

8.2.2.1 Score Propagation

Score propagation allows us to obtain scores from different layers of propagators. k—Deep social scores get the actor's score in the presence of itself. If not, it asks the actor's friends to provide their scores using their friends' scores to predict their scores. If no one returns a score, the score is unpredictable. Some improvements have been down to Score Propagation. Figure 8.2 shows the Score Propagation Network without Score.

1. The idea of similarity, which is comparable to weighting the propagated scores, is introduced by correlation. Furthermore, the default score option enhances the system's unpredictability, breaking the limits of the old trust system and allowing users to explore more freely. If the user does not rate the item, the default score is calculated and returned with a specified likelihood. The user's average score and the average score of the item are the two default approaches (Meyffret et al. 2012a, b).
2. The Confidence idea was developed in response to an issue in which the average weighted weight does not function when there is just one friend. Instead, the confidence is sent along with the score, and after aggregation, the confidence is multiplied by the current weights to create a new weight to determine the final score. Figure 8.3 shows the Score Propagation with Weight Values.

Fig. 8.2 Score propagation network without score

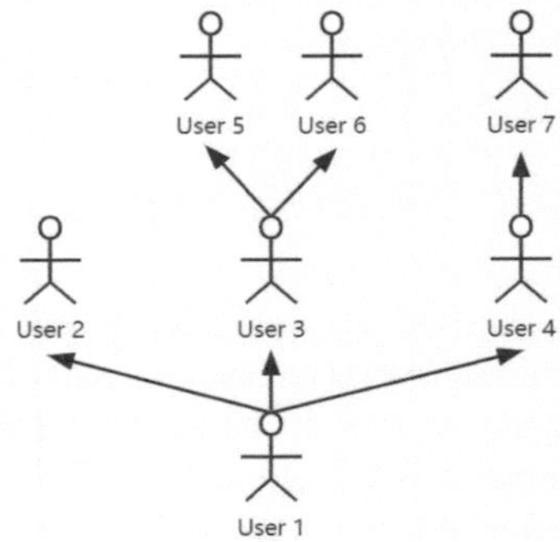

Fig. 8.3 Score propagation with weight values

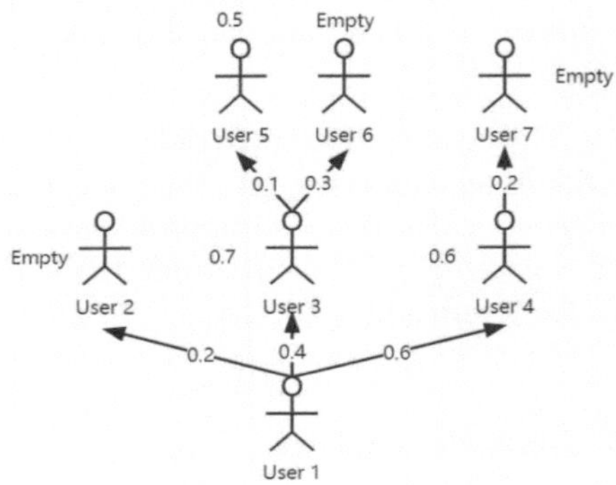

8.2.3 *Evolution of Trust-Based Recommender Systems*

The system evaluates the fitness by Coverage, Root Mean Square Error (RMSE), and F-Measure is the statistical measurements. Equations 8.4, 8.5, and 8.6 show the calculation of RMSE and F-Measure. The coverage is the percentage of anticipated ratings out of all possible ratings. It doesn't say if the ratings were predicted accurately, but it does highlight how many forecasts an algorithm can make. The RMSE stands for the prediction's average error. It's essentially the standard deviation of the error without the mean. The RMSE is a measure of how accurate a forecast is. The lower the RMSE, the more accurate the prediction (Meyffret et al. 2012a, b).

The paper uses the classical leave-one-out method for the evaluation of the dataset, i.e., for the whole dataset, removing one of the ratings at a time and trying to predict it using other data.

$$\text{RMSE} = \sqrt{\frac{\sum_{n=1}^{N}(p_n - r_n)^2}{N}} \tag{8.4}$$

$$\text{Precision} = 1 - \frac{\text{RMSE}}{\text{range}} \tag{8.5}$$

$$F_1 = \frac{2 * \text{Precision} * \text{Coverage}}{\text{Precision} + \text{Coverage}} \tag{8.6}$$

Range indicates the value range of the score.

8.2.4 Concept of Attack in Recommender System

While the performances of recommender systems can be affected by their intrinsic defects, there are some extrinsic disturbance factors. For example, some noisy data may be generated by the users if they willingly give biased ratings. Besides these, attacks can exist in recommender systems. A shilling attack, which can also be referred to as a profile injection attack or data poisoning attack, can easily influence the performance of a collaborative filtering recommender system even if only a small number of malicious profiles are injected into the system (Williams et al. 2007). An attack can be carried out by injecting carefully crafted profiles to achieve the desired purpose of either promoting or demoting a specific item. The attack for promotion is called a push attack while the one for demotion is called a nuke attack. Numerous pieces of research have investigated different attack strategies and attack detection techniques.

8.2.4.1 Attack Strategy

O'Mahony et al. (2005) divided the task of building attack profiles into two sub-tasks. The first concerns the selection of items that together with the target item constitute the profile and the second relates to the ratings given to the selected items. This group proposed two attack strategies which are popular attack and probe attack. Popular attack selects popular items which are prevalently liked or disliked by the public. By choosing the popular items to build the attack profiles, the cost of an attack can be minimized since such a profile is highly possible to be located in the neighborhood of many genuine users. A probe attack involves probing the recommender system and selecting the items based on the recommendations provided by the system. This type of attack requires less knowledge about the system and is easier to implement. Only a small number of items should be selected by the attacker as the initial seed to derive recommendations from the system and attack profiles can be created according to the recommendations progressively. Profiles created by probe attacks tend to have high similarity with genuine users and are difficult to distinguish from genuine users.

Burke et al. (2006) investigated the attack strategies more meticulously. They divided the attack profile into four categories, namely selected items, filler items, unrated items, and the target item. Different attack models treat these items in different ways. These attack models include random attack, average attack, bandwagon attack, segment attack, and love/hate attack. The concrete rating methods for some of these attack models are provided in Cao et al. (2013). Aggarwal (2016) explained these attacks in more detail and their attack models involve null items, filler items, and the target item, which are slightly different from the above two research groups. The following are some concrete explanations for these different types of attacks. Table 8.1 describes the generation methods for filler items for these attack models.

- Random Attack—Filler items are selected randomly and ratings complying with a probability distribution around the global mean are assigned to the filler items. This type of attack requires the knowledge of the global mean which is the mean value of all ratings across all items.
- Average Attack—Filler items are selected randomly and for each specific selected item, the average value of ratings on this item given by other users are assigned in the attack profile.
- Bandwagon Attack—Filler items contain two parts. The first part consists of popular items which are widely liked and are assigned the maximum allowed rating value. The second part incorporates randomly selected items and is rated randomly. The type of attack does not require knowledge of the rating matrix but the attacker needs to know what items are popularly liked by the users.

Table 8.1 Filler items selection and rating rules for different attack models

Attack model	Generation method for filler items
Random attack	Global mean rating $\rightarrow$ randomly selected items
Average attack	Average rating value of the specific item $\rightarrow$ randomly selected items
Bandwagon attack	$r_{\max} \rightarrow$ popularly liked items Random rating $\rightarrow$ randomly selected items
Popular attack (push)	$r_{\min} \rightarrow$ widely disliked items $r_{\min} + 1 \rightarrow$ widely liked items
Popular attack (nuke)	$r_{\max} \rightarrow$ widely liked items $r_{\max} - 1 \rightarrow$ widely disliked items
Love/Hate attack	$r_{\min} \rightarrow$ nuked item $r_{\max} \rightarrow$ other items
Reverse bandwagon attack	$r_{\min} \rightarrow$ widely disliked items
Probe attack	$r_{\max} \rightarrow$ items recommended by the system based on the seed profile
Segment attack	$r_{\max} \rightarrow$ the pushed item and the items of the same segment (category)

- Reverse Bandwagon Attack—This type of attack is the reverse of a bandwagon attack and is designed to demote items. Instead of using widely liked items, the reverse bandwagon attack uses widely dislike items as the filler items, which have received many low ratings.
- Love/Hate Attack—This type of attack is designed to be a nuke attack. The minimum allowed rating value is assigned to the target item while the maximum allowed rating value is assigned to the other items.
- Segment Attack—While most of the other types of attacks are ineffective on item-based recommender systems, segment attacks perform effectively on this type of recommender system. In a segment attack, a set of items that are considered to be in the same segment (category or genre) as the target item are selected. These items are assigned the maximum allowed rating value together with the target item. Another set of sampled filler items is assigned the minimum allowed rating value. The purpose is to maximize the variations of the similarities among items in the same segment, which gives an advantage to the target item.

The assignment of a rating for the target item is dependent on whether the attack wants to promote the item (in a push attack) or demote the item (in a nuke attack), apart from the love/hate attack and reverse bandwagon attack which is designed to be nuke attacks and the segment attack which is designed to be push attacks. To promote the item, the maximum allowed rating value will be given. To demote the item, the minimum allowed rating value will be given. Aggarwal (2016) also introduces popular attacks and probe attacks which have already been discussed.

While most of the previous research focused on the above-mentioned attack models, some research groups claimed that when attacking a recommender system, the attacker may not always have enough knowledge about the system to perform these attacks. Fan et al. (2021) proposed a framework called CopyAttack, and Song et al. (2020) proposed a framework called PoisonRec, both of which are black-box attacking strategies based on reinforcement learning. These two frameworks require no knowledge about the recommender system and may be more feasible in real scenarios.

8.2.4.2 Attack Detection Technique

Attack detection aims to distinguish between attack profiles and genuine profiles. Therefore, attack detection can be treated as a classification problem. Classification features can be extracted from the rating matrix-like Rating Deviation from Mean Agreement (RDMA), Weighted Deviation from Mean Agreement (WDMA) (Burke et al. 2006), etc. Typical solutions for a classification problem include Bayes learning, K-nearest neighbors (KNN), decision tree, support vector machine (SVM), etc. Williams et al. (2007) compared the performances of KNN, C45, and SVM for attack detection and concluded that SVM adds significant robustness to the recommender system. Cao et al. (2013) proposed Semi-SAD which is a semi-supervised learning-based shilling attack detection algorithm. This algorithm first trains a naive

Bayes classifier on a small set of labeled users and then incorporates unlabeled users into the classifier. Two more algorithms were used for performance comparison. The first was a supervised naive Bayes classifier called Bayes-SAD, and the second was an unsupervised approach based on a principal component analysis called PCA-SAD. The result of this research demonstrated that Semi-SAD can better detect various kinds of shilling attacks than the other two algorithms.

8.3 Challenges in Previous Research

8.3.1 Challenges in Researches About Trust

The score propagation system introduces randomness and can provide novelty recommendations for specific users, but maintains a low frequency to prevent noise. However, this randomness cannot continuously converge or disperse after a change in user preferences. For example, a specific group of fan users may be interested in only a very small fraction of the content. A more ideal model should have novelty recommendation results converge gradually as users engage in system interactions until they remain at a low frequency. After the user's interest changes, it gradually diverges and increases the novelty recommendation content nonlinearly.

8.3.2 Challenges in Researches About Attack

Previous research made many assumptions about the attacks with no appropriate proof. For example, Aggarwal (2016) claimed that attack profiles tend to have high self-similarity and therefore a cluster with a smaller radius is considered the attack cluster. However, as the number of users is surging, the aforementioned assumption may not always be true. If the number of users is larger enough, it may be inevitable that an unexpected number of genuine users are included in the cluster which is considered the attack cluster, which makes it more difficult to distinguish between attackers and genuine users. Attack profiles and real profiles can be mixed in the same cluster, and there is no proof that the proportion of attack profiles in a cluster with a smaller radius is high.

Another assumption made by many research groups which may not always be practical is that the attackers are allowed to create abundant accounts to perform profile injection attacks. This assumption may hold when there is no restriction on account creation for most of the applications. However, since more and more applications impose rigorous restrictions on account creation, especially in China, this assumption will fail in many cases. Real name registration becomes compulsory for many applications in China, especially for e-commerce applications, which means the users must register with their real name and ID number to activate their account, otherwise they are not allowed to do a transaction or even enter the application. In this scenario, profile injection attacks or shilling attacks become impossible, and as a consequence, it is not always necessary to defend the recommender system against these attacks by detecting and removing fake profiles when such administrative measures are enforced.

Apart from the impractical assumptions, the evolution of attacks is also making the attack-resistant mechanisms less effective. While most of the previous research focused on traditional attacks like random attacks, average attacks, etc., new types of attacks are appearing. Some genuine users can be utilized by the attacker to perform the attack. Specifically, the attacker may distribute some money or give some other benefits to a group of genuine users and hire them to give ratings to the target item. In this way, these genuine users perform the attacks under the guidance of the real attacker, and there will be no clear boundary between a fake profile and a genuine profile, which means fake ratings may exist in a genuine user's profile. This type of attack requires no knowledge about the recommender system since the attacker does not need to create profiles. It can be easily implemented with some economic cost and can be highly effective. If detection has to be performed, every single rating in the rating matrix should be investigated and the ratings should be categorized into genuine ratings and fake ratings, which is almost impossible to be done. Therefore, some countermeasures other than detecting and removing the attack profiles or attack ratings can be designed to improve the performance of the recommender system and minimize the effect of the attacks.

Another defect is that most of the previous research only considered the rating matrix and the recommender algorithm when creating the attack profiles or doing the detection. However, many other factors can be taken into account. For example, a recommender system can be time-sensitive and make use of the time at which a rating is given to analyze a user's behaviors.

In conclusion, the complexity of the attack strategies and lack of a comprehensive perspective become a big challenge for the relevant research topics. The attack strategies and defense strategies are improving in an adversarial way. While more robust defense strategies are being developed, researchers should also learn about new attack strategies to ensure the effectiveness of their pro- posed defense strategies. In the meantime, the assumptions made to support the research should adapt to the real scenarios which tend to be mutable.

8.4 Possible Improvements for Future Research

8.4.1 Improvements in Score Propagation Model

The following two strategies show the divergence method and the convergence method, respectively. Convergence indicates that users will receive increasingly accurate tweets, while the divergent approach ensures that users have access to certain other domains of information.

8.4.1.1 Overload Weighted Weight

In the score propagation model, when we use weighted weight to determine the influence of a user there is an upper bound to avoid enormous effects. However, in some specific areas like fan groups and academic forums, users are specific groups and most likely do not need diverse project recommendations. Gradually converging items and information may be more in line with demand.

In this case, allowing the weights to exceed the upper limit will lead to better results. Extended analysis of user behavior means user feedback was collected implicitly and explicitly utilizing click-stream and user activity data for our RS extension with weighted multi-attributes (Akcayol et al. 2018). After this, the system can identify accurate user profiles and thus feel that those users can be overloaded with the weight of the item, thus obtaining progressively converging recommendation results.

However, to avoid convergence to a minimum, any should set a small random number to generate recommendations for items that are not related to the domain. When a user's click event occurs, it means that the user has become interested in items outside this domain, in which case we should revert the weights to the upper limit.

8.4.1.2 A Trust-Based Model Using Simulated Annealing Meta-heuristic

Inspired by a semantic-based recommender system using a simulated annealing algorithm (Picot-Clémente et al. 2010), this paper proposes a new trust-based recommender system model. The workflow is shown in Fig. 8.4.

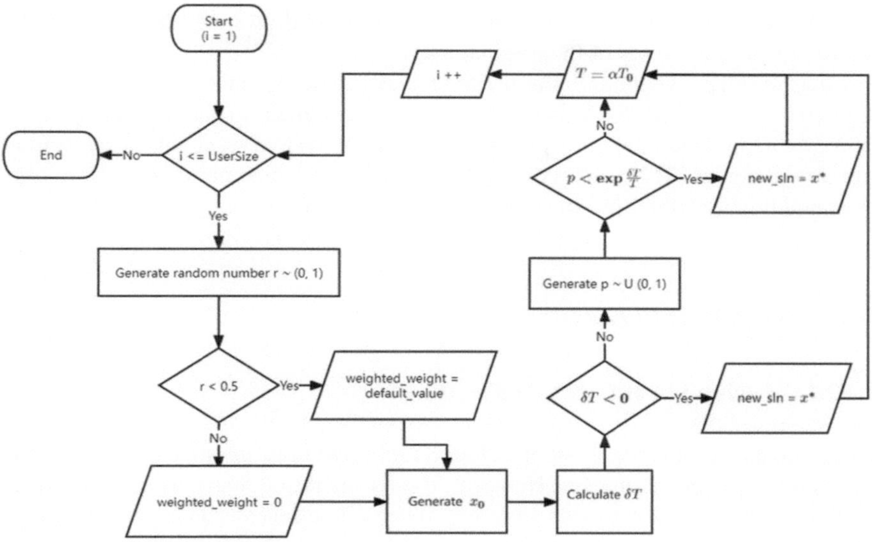

Fig. 8.4 SA procedure diagram

Algorithm 1 SA

Input: T_0, α, x_0 and *UserSize*
Output: BestSln x^*
function Simulated Annealing (*defaultValue*)
  for $i = 1 \rightarrow UserSize$ **do**
    Generate $r\ U{\sim}(0, 1)$
    if $r < 0.5$ **then**
      Use default weighted weight generates x_0
    else
      Use 0 *as weighted weight generates* x_0
    end if
    Calculate δT
    if $\delta T < 0$ **then**
      Accept x^* as the new solution
    else
      Generate $p \simeq U(0, 1)$
      if $p < \exp^{\delta T_T}$ **then**
        Accept x^* as the new solution
      end if
    end if
  end for
  $T = \alpha T_0$
end function

For this algorithmic model, a solution is equivalent to a complete trust network. In this, each layer of the network consists of several users. The first layer is the original user, the second layer is then the friends of the original user, and the third layer is the friends of the friends of the original user.

Each layer has a weighted value to the next layer. Root Mean Square Error (RMSE) is chosen to be the fitness value that reflects the accuracy of the system. With the system at a higher temperature, the probability of the system receiving a less accurate solution increases. This causes the system to try different combinations of weights, thus avoiding the system from falling into a local optimum.

8.4.2 User's Activeness

8.4.2.1 Evaluation of User's Activeness

To defend a recommender system against attacks, the most intuitive way is to detect and remove the attack profiles. However, if only the rating matrix is used for attack profile detection, the detection may not always be effective since the attacks are becoming more and more complex. Therefore, some other features derived from user behaviors can be used when designing the recommender algorithms and some countermeasures other than detection and removal can be taken to mitigate the effect of attacks and increase the cost to perform attacks.

One factor which is neglected by many research groups is the time at which a rating is given. By investigating the distribution of time at which a specific user gives ratings to different items (which will be referred to as rating time in the following contents) and the total number of ratings given by this user, the activeness of this user can be analyzed, which can be used as a feature when designing the recommender algorithms.

One feasible way we propose to quantify a user's activeness makes use of the information entropy contained by the distribution of the rating times of this specific user. Information entropy is a measure of histogram dispersion. The distribution of the rating times of a specific user can be described by a histogram, which contains bins separated on a daily, weekly, or monthly basis. Every bin represents the number of ratings given by this specific user within the period of that bin. The entropy can be calculated as Eq. 8.7:

$$H(X) = -\sum_{k=0}^{L-1}\big(p(r_k)\log_2 p(r_k)\big) \tag{8.7}$$

where $H(X)$ is the information entropy of the distribution of the rating times of a specific user, L is the number of bins contained by the histogram, and $p(r_k)$ is the possibility of a rating time belonging to the kth bin. $p(r_k)$ can be calculated as the quotient of the sum of the values of all bins divided by the value of the kth bin (with the starting index of 0).

A more dispersed histogram will have a larger entropy and a less dispersed one will have a smaller entropy. When all the instances fall in one single bin and all the other bins are empty, the entropy for this histogram will become 0. When the instances fall uniformly in every bin, which means each bin has the same value, the entropy of the histogram will reach the maximum value. To be more detailed, when a histogram on a daily basis containing 7 days is used to calculate the entropy, a user who only gives ratings on one single day but becomes inactive on other days, will have an entropy of $-1 * \log_2 1 = 0$ (in this case, L is 1, and $p(r_k)$ will only have one value which is 1). If a user gives the same number of ratings every day, the entropy will become $-7 * \frac{1}{7} * \log_2 \frac{1}{2} = 2.81$.

Figure 8.5 is an example of the histogram of the distribution of rating times of a single user over one week on a daily basis. In this specific case, the entropy is calculated as $-0.32 * \log_2 0.32 - 0.12 * \log_2 0.12 - 0.16 * \log_2 0.16 - 0.24 * \log_2 0.24 - 0 - 0.12 * \log_2 0.12 - 0.04 * \log_2 0.04 = 2.36$.

The entropy only considers the possibilities of the rating time distributions but neglects the quantity. Therefore, apart from the entropy which reflects the dispersion of a user's rating times, another two factors should also be considered to quantify the user's activeness, which is the total number of ratings given by this specific user and the average (or median) number of ratings given by some or all the other users. We divide the user's activeness further into two types, global activeness $\left(\mathrm{act}_{\mathrm{global}}(u)\right)$, and local activeness $(\mathrm{act}_{\mathrm{local}}(u))$, which are respectively defined as Eq. 8.8:

$$\mathrm{act}_{\mathrm{global}}(u) = \frac{n_u}{n(u)} \cdot H(R_u) \tag{8.8}$$

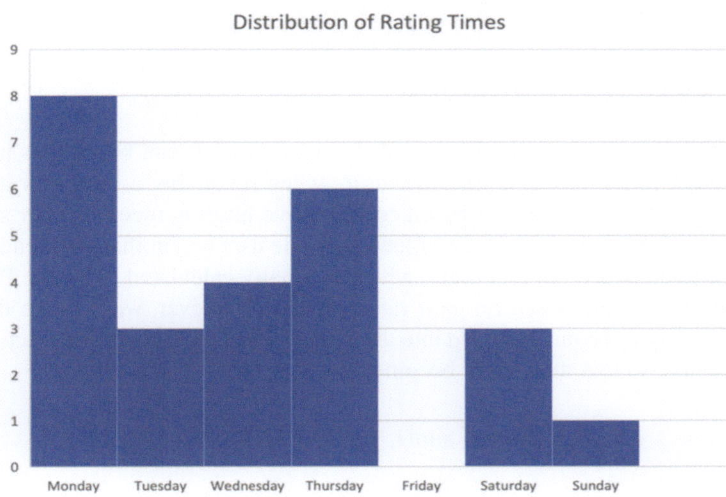

Fig. 8.5 An example of the histogram of the distribution of rating times of a single user over one week on a daily basis

where n_u is the total number of ratings given by user u, $n(\overline{u})$ is the average number (which can be substituted with a median number) of ratings given by his near registered users who register their accounts on the same day or in the same week or month as this specific user, and $H(R_u)$ is the information entropy of all the rating times of user u on a basis of the same period as the former measure, and Eq. 8.9:

$$\text{act}_{\text{local}}(u) = \frac{n_u}{\overline{n}} \cdot H(R_u) \tag{8.9}$$

where n_u is the number of ratings given by user u within a specific period such as a month, $\overline{n}$ is the average number (which can be substituted with a median number) of ratings given by all the other users within the same month, $H(R_u)$ us the information entropy of the rating times of user u over this period on a daily or weekly basis.

For the local activeness, the whole period of the histogram and the span of a single bin can vary if necessary. Local activeness may be preferred when considering a user's activeness as a feature in the recommender algorithms since recent data tend to have higher reference values. In addition, the global activeness may be slightly biased due to the following two reasons. Firstly, the concrete registration time for a specific user tends to be slightly different from his near registered users but they are categorized into the same group and are considered to have registered in the same period. Secondly, near registered users from different groups may not have the same number of members, and groups registered in different periods may not always have similar behaviors. However, if we ignore the slight discrepancies in the registration times of the users in the same group and assume that the numbers of the members in every near registered group are the same and they behave quite similarly, the global activeness can be considered unbiased.

8.4.2.2 Usages of User's Activeness

Users' activeness can be used in several different ways. It can be used as a factor of the weight of the influence of other users when predicting a rating for a target user on a specific item in a user-based collaborative filtering recommender system. In such a system, the weight of the influence of one user on another user tends to be the similarity between the two users, which is usually calculated using the Pearson correlation coefficient based on their rating matrix. However, we propose that the activeness of user j be incorporated into the weight. Here we take the local activeness as the example so that the weight becomes Eq. 8.10:

$$w_{i,j} = \text{sim}(i, j) \cdot \frac{n_j}{\overline{n}} \cdot H\left(R_j\right) \tag{8.10}$$

where sim(i, j) is the similarity between user i and user j. However, such a weight will be dominated by some extraordinarily active users. Therefore, some variation of the rectified linear unit can be utilized to mitigate the effects of such users. After applying the rectification, the weight becomes Eq. 8.11:

$$w_{i,j} = \text{sim}(i, j) \cdot f\left(\frac{n_j}{\overline{n}}\right) \cdot H\left(R_j\right) \tag{8.11}$$

where f is a variation of the rectified linear unit, which is defined in $[0, +\infty)$ and has the form of Eq. 8.12:

$$f(x) = \begin{cases} x, \text{ if } x < \alpha \\ \alpha, \text{ if } x \geq \alpha \end{cases} \tag{8.12}$$

where α is some predefined value to prevent the dominance of extraordinarily active users and should be at least larger than 1. The value of α can be a decimal and we propose that the value of α should be less than 3, which means that a user is allowed to have at most three times the effect of an average user with the same entropy.

When the user's activeness is taken into consideration and combined into the calculation of the weights, the effect of attacks will be greatly mitigated. Users who give ratings frequently on different days with a sufficient quantity will be considered active and these users will dominate the recommender process in a user-based collaborative recommender system while the new users or inactive genuine users will only play a modest role. As for the attackers who try to inject abundant fake profiles in one day, they will completely fail the attacks since these quickly injected profiles will be assigned activeness of 0, which leads to a weight of 0 in the recommender process. These fake profiles will never be trusted by the recommender system. Therefore, if an attack needs to succeed, the attacker should be active for a long time period and give enough ratings in order to obtain the trust from the system.

The above-proposed calculation method for weight can also be applied when calculating the default weight in the Simulated Annealing algorithm proposed in Sect. 8.4.1.2.

Besides incorporating the user's activeness into the weight, the user's activeness may also be useful if treated as a classification feature when doing an attack detection using machine learning techniques.

8.4.3 Administrative Measures

8.4.3.1 Real Name Registration

Apart from the above-mentioned strategies, some administrative measures can also be important to enhance the robustness of the recommender system. For example, real name registration can be one feasible proposal that has been discussed in Sect. 8.4.2. If real name registration should be implemented, the developers should not only defend the system against recommender system attacks, but they also need to prevent network attacks to protect users' privacy. Therefore, some more complicated technical measures and administrative measures should be designed to ensure server security and protect users' sensitive personal information, which should be compulsory for real-name registration. The following sections briefly describe some administrative measures concerning server security, which may pave the way for real-name registration.

Password Policy

A good password policy should be formulated to protect the user's account, which contributes to the security of the user's personal information. Passwords tend to be stored in a hashed format using one-way functions rather than in plaintext and therefore the passwords become unreadable. However, it is possible to crack the hashed passwords to recover the original passwords. Brute force attack and dictionary attack are two typical offline password cracking strategies. Such offline attacks utilize leaked hashed password files and try numerous character combinations to see whether there is a hash collision. A brute force attack tries an extensive number of possible passwords and can be well supported by GPUs with fast computational capabilities. However, brute force attacks are inefficient to crack long passwords. A dictionary attack tries different combinations from several dictionaries as possible passwords. A dictionary can be a general dictionary or a special one that may contain a set of leaked plaintext passwords or user information such as names, birthdays, etc. The following is an example password policy that can prevent these attacks to some extent: The length of the password should be between 8 and 16.

- At least one lowercase letter, one uppercase letter, one punctuation, and one numeric number should be contained.
- No dictionary words or keyboard sequences should be used.
- No personal information should be contained such as phone number, name, birthday, etc.
- The password should be changed on at least a semiannual basis.
- The password should not be changed to a previously used one.

Additional Authentication Measures

Apart from the password policy, some additional measures can be implemented for further authentication. Two-step authentication can be used to enhance account security. The first step is to check the user-defined static password which may follow the above rules. A one-time password can be utilized for the second step and it can be implemented in different ways. For example, many companies use short message services to provide users with a PIN code through cellphone communication. Hardware tokens such as microprocessor-based smart cards or pocket-size key fobs can also be used to generate a one-time password. With the two-step authentication, the user's account will be much more secure.

Device lock can be another authentication measure. With a device lock, the users can only log in to their accounts on authorized devices. The users are required to bind their accounts with their phone numbers or emails. Then, when a user needs to log in to his account on a new device, a PIN code will be sent through a short message service on the phone or sent via email. In this way, even if the passwords are leaked, the users' accounts remain secure.

8.4.3.2 IP Address Monitoring

IP address monitoring can also be one administrative measure to defend the system against recommender system attacks. When an abnormal number of profiles are created by the same IP address, these profiles may be created by an attacker. Therefore, these suspicious profiles can be removed from the system and a blacklist firewall can be constructed to drop the packets from this IP address.

8.5 Summary

In this chapter, we introduce the concept of trust and attack in recommender systems. Specifically, we describe the definitions, properties, and some models of trust, and we explain how social scoring can be applied in recommender systems. We also briefly talk about the evolution of trust-based recommender systems. As for the concept of attack, some widely existing attack strategies and some innovative strategies are discussed, and we do a brief survey about the attack detection techniques. Furthermore, some challenges in previous research about trust and attack in recommender systems are pinpointed, and according to these challenges, several possible solutions are proposed. The improvements in the score propagation model, the definition and usage of user's activeness, and some administrative strategies are proposed to enhance the performance of recommender systems, which can be more reliable and resilient. In conclusion, although the accuracy of recommendations given by a recommender system is of paramount importance, there are other heterogeneous factors such as trust and attack which should not be neglected when designing an effective and robust recommender system.

Think Tank

1. *What is social scoring?*
2. *What is trust in a recommender system and what are its properties?*
3. *What is score propagation?*
4. *What are the different types of attack methods?*
4. *What are the challenges in research in attacks?*

References

Abdul-Rahman A, Hailes S (1998) A distributed trust model. In: Aggarwal CC (ed) New security paradigms workshop. Recommender systems, vol 1. Springer, Berlin

Aggarwal CC (2016) Recommender systems, vol. 1. Springer

Akcayol MA, Utku A, Aydoğan E, Mutlu B (2018) A weighted multi-attribute-based recommender system using extended user behavior analysis. Electron Commer Res Appl 28:86–93. Retrieved from https://www.sciencedirect.com/science/article/pii/S1567422318300164; https://doi.org/10.1016/j.elerap.2018.01.013

Burke R, Mobasher B, Williams C, Bhaumik R (2006) Classification features for attack detection in collaborative recommender systems. In: Proceedings of the 12th ACM SIGKDD international conference on knowledge discovery and data mining, pp. 542–547

Cao J, Wu Z, Mao B, Zhang Y (2013) Shilling attack detection utilizing semi-supervised learning method for collaborative recommender system. World Wide Web 16(5):729–748

Fan W, Derr T, Zhao X, Ma Y, Liu H, Wang, J, Tang J, Li Q (2021) Attacking black-box recommendations via copying cross-domain user profiles. In: 2021 IEEE 37th international conference on data engineering (ICDE), pp 1583–1594. https://doi.org/10.1109/ICDE51399.2021.00140

Golbeck J (2006) Generating predictive movie recommendations from trust in social networks. In: Stølen K, Winsborough WH, Martinelli F, Massacci F (eds) Trust management. Springer, Berlin, pp 93–104

Golbeck JA, Hendler J (2005) Computing and applying trust in web-based social networks. PhD dissertation, University of Maryland at College Park, USA (AAI3178583)

Haydar C (2014) Les systèmes de recommandation à base de confiance (trust-based recommender systems)

Haydar C, Roussanaly A, Boyer A (2013, Nov) Individual opinions versus collective opinions in trust modelling. In: SOTICS 2013, the third international conference on social eco-informatics, Lisbon, Portugal, pp 92–99. Retrieved from https://hal.inria.fr/hal-00929925

Marsh SP (1994) Formalising trust as a computational concept

Massa P, Bhattacharjee B (2004) Using trust in recommender systems: an experimental analysis. In: Proceedings of itrust2004 international conference, pp 221–235

Meyffret S, Médini L, Laforest F (2012a) Trust-based local and social recommendation. Association for Computing Machinery, New York, NY, USA. Retrieved from https://doi.org/10.1145/2365934.2365945

Meyffret S, Médini L, Laforest F (2012b, 04) Recommandation sociale et locale basée sur la confiance. Doc numérique 15:33–56. https://doi.org/10.3166/dn.15.1.33-56

O'Mahony MP, Hurley NJ, Silvestre GC (2005) Recommender systems: attack types and strategies. In: AAAI, pp 334–339

Picot-Clémente R, Cruz C, Nicolle C (2010, Oct) A semantic-based recommender system using a simulated annealing algorithm

Selmi A, Brahmi Z, Gammoudi MM (2016) Trust-based recommender systems: an overview. In: Proceedings of 27th international business information management association (IBIMA) conference, Milan, Italy

Song J, Li Z, Hu Z, Wu Y, Li Z, Li J, Gao J (2020) PoisonRec: an adaptive data poisoning framework for attacking black-box recommender systems. In: 2020 IEEE 36th international conference on data engineering (ICDE), pp 157–168. https://doi.org/10.1109/ICDE48307.2020.00021

Sridevi M, Rao RR, Rao MV (2016) A survey on recommender system. Int J Comput Sci Inf Secur 14(5):265

Williams CA, Mobasher B, Burke R (2007) Defending recommender systems: detection of profile injection attacks. SOCA 1(3):157–170

Chapter 9
Steps in Building a Recommendation Engine

Abstract In this chapter, we discuss the steps one needs to keep in mind while designing an efficient recommender system. We also see what are the design parameters for rating the efficiency of a recommender system. Then the steps to build such a system are discussed along with a generic architecture.

Keywords Design · Efficiency · Evaluation parameters · General structure of a RS · Classification model

9.1 Introduction

A recommendation engine (sometimes referred to as a recommender system) is a tool that lets algorithm developers predict what a user may or may not like among a list of given items (Gediminas and Alexander 2005; Guibing et al. 2013; Breese et al. 1998). Earlier people used to purchase products based on the recommendations of their close groups of trustworthy family or friends. However, in more recent times, with an increase in the variety of products to choose from and with the advent of digital platforms, there is also a huge increase in the volumes and variety of data and information that needs to be processed and refined before one can arrive at a choice. So users are now relying more and more on recommendation engines to help them in making the best choices. Therefore, building a robust and accurate recommendation engine is the first and most important step to enabling service providers to provide the most relevant and accurate suggestions or recommendations of products to their clients.

However, before the design phase, we need to consider the evaluation parameters of a recommender system, which are accuracy, coverage, confidence and trust, novelty, serendipity, diversity, robustness and stability, and scalability. The rest of the chapter is organized as follows: Sect. 9.2 discusses the evaluation parameters of designing an efficient recommendation system, Sect. 9.3 gives an overview of the

P. Kar et al., *Recommender Systems: Algorithms and their Applications*, Transactions on Computer Systems and Networks, https://doi.org/10.1007/978-981-97-0538-2_9

ways of designing an efficient system, Sect. 9.4 discusses the steps to design a recommendation engine, Sect. 9.5 discusses the architecture, and finally the summary in Sect. 9.6.

9.2 Design/Evaluation Parameters of a Recommender System

All predictive models or recommender systems rely very heavily on large volumes of data. The more data a machine learning algorithm has, the better will be the results it returns. It is because of this reason that the organizations which have the best recommender systems are those which have access to huge volumes of data, like Amazon, Google, and Netflix. In this section, we discuss the main parameters for evaluating the efficiency of a recommender system. While some of the parameters can be quantified concretely, some other parameters are more subjective and rely on the user (Fig. 9.1). The factors are as follows:

a. **Accuracy**—Accuracy is one of the most important parameters for measuring the efficacy of a recommender system. Since most of the ratings are numerical data, so the metrics for accuracy are similar to those used in regression modeling. There can be errors at various stages of the system. If a rating matrix R contains a known user rating r(act), then if a recommendation system estimates this rating as r(est), the entry-specific error of the system will be r(est)-r(ct). The overall errors also may be calculated using various methods. Moreover, all the entries for a rating matrix cannot be used for training the model or for accuracy metrics as it will lead to overfitting. However, for a recommender system to be successful, it's not all about accuracy. We need to consider many other performance metrics and not just the accuracy only. Focusing only on the accuracy aspect can actually be detrimental to the system. Suppose a person is using a travel recommendation system. If the system returns with recommendations of places that have already been visited by that person, then it would probably be rated as a poor recommendation system.

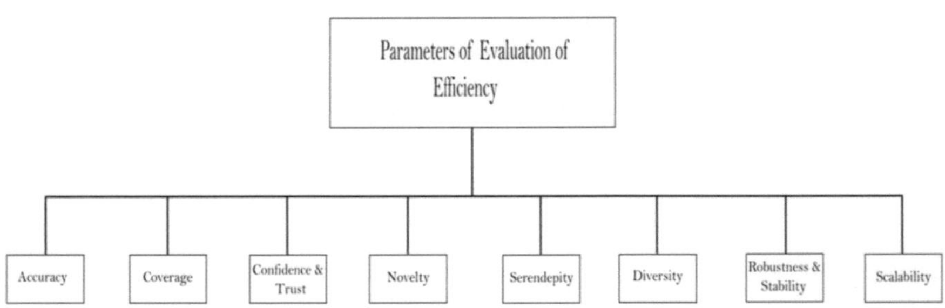

Fig. 9.1 Parameters for evaluating the efficiency of a recommender system

So basically the accuracy parameter tells if the recommender system is able to predict those items that a person has already rated or interacted with. Therefore to develop a balanced system, other parameters also need to be taken into consideration.

b. **Coverage**—Even when a recommendation system is highly accurate, it may not ever be able to recommend a particular proportion of items, or ever recommend to a certain portion of users. This is called the coverage and is caused by the fact that some rating matrices can be sparse, containing a single entry for each row and each column. Under such circumstances, no meaningful recommendations can be made, no matter which algorithm is used. But still, various recommender systems have varying levels of bounds for providing coverage. So usually in practical settings, the systems mostly have 100% coverage, because they use default ratings in place of ratings that are not possible to be predicted. For e.g. default ratings may be used as the average of all the ratings of a user for an item, when the specific user-item combination cant be predicted. So there has to be a trade-off between the accuracy and the coverage of a recommendation system.

c. **Confidence and Trust**—Estimating the ratings is not an exact process, because the ratings can vary significantly based on the training data that is used. Apart from this, the algorithmic logic used may also impact the ratings to a large extent (Jennifer 2005; Paolo and Paolo 2007; Roger et al. 1995). So this leads to uncertainty in the users regarding the accuracy of the predictions made. So most recommender systems have reports along with confidence estimates, i.e. there is a confidence interval (usually 95% or 99%) provided along with the rating predictions. Usually, users tend to trust recommender systems with smaller confidence intervals. But it is only possible to compare two algorithms if they have the same confidence intervals, but if they have different confidence intervals then they cannot be compared meaningfully. So to compare any two systems, we need to set the confidence level of both of them to be the same.

Confidence is the measurement of the faith in the system on its recommendation, while trust is a measurement of the user's faith in the system. So basically the trust level measures the level of faith that a user has in the recommendations provided to him/her. This is a very important factor because even if the ratings provided by a system are accurate, if the user has little trust in the system, then it is of no use. So trust is closely related to accuracy but they are not quite the same thing. It is also not the same as usefulness. e.g. if a system suggests some items to a user which are already liked and known to him/her, then although it increases the user's trust in the system, it is not really very useful to the user. This goal also contradicts the aim of the novelty parameter, where recommendations already known to the user are usually discouraged. So while designing a recommender system, it is very common to trade off one parameter against the other.

Generally, the most prevalent way to measure trust is to conduct user surveys, where the users can explicitly share their ratings or trust levels of a system. There are at present many online evaluation surveys present, as it is more difficult to measure trust through offline surveys.

d. **Novelty**—The novelty of a recommender system defines the probability of the system providing options and choices to the users, that they are unaware of or unknown to them. If a user gets unseen recommendations, then it helps the user to discover important information about their preferences that they were previously not aware of, which is probably more useful than the discovery of choices that they already know but have not rated yet. As we saw previously, a small number of recommendations may be useful in improving the trust of a user in the system, but it might not go a very long way in improving the rates of conversion. So the most commonly used method for the measurement of the novelty of a system is to conduct online experiments where users are asked about their prior familiarity with an item. Novel recommender systems are usually expected to recommend items that have a higher probability to be selected in the future as compared to the present. So all ratings that have been created after a particular time say t1 and some ratings before the time t1, will be removed from the training data, after which the system is trained. Now the items which have been removed are used for scoring. So each of the items which have been rated before time t1 and have been recommended correctly will be penalized. Conversely, every item that has been rated after time t1 and recommended correctly will be rewarded. The basic idea here is that this method will create a differential accuracy between the past and future predictions, the assumption being that if an item is popular then it is less likely to be novel, so less credit will be awarded for recommending items that are more popular.
e. **Serendipity**—The term serendipity means a "lucky discovery". For recommender systems, it actually measures the level of surprise for successful recommendations, which is slightly different from the novelty parameter. In fact, it is a slightly stronger condition than novelty. While the requirement for the novelty factor is that the user is only unaware of the recommendation, the serendipity factor needs the recommendation to be unexpected. For example, if a person normally eats Thai or continental food, then the novelty condition will prompt the system to recommend new Thai or continental food restaurants near him that the person is not aware of. On the other hand, the serendipity factor may suggest an Indian restaurant because it is a less obvious choice and has a surprise element to it.
f. **Diversity**—Diversity in a recommendation system means that the set of proposed recommendations in a single recommended list should be as different from each other as possible, e.g., a recommendation list of the top 3 movies is sent to the user. Then if all the movies are in the same genre and have similar casts, it is not a very good system because if a viewer rejects the first choice, then there is a very high chance that the user will reject the other two movies listings also. So if different types of movies are presented then there is a higher chance that the user may choose a movie from the list. It is to be noted however that diversity of a system is always measured over a particular set of recommendations. A system with an increase in diversity usually has increased novelty and serendipity also.

g. **Robustness and Stability**—A recommender system should also be robust and stable enough to withstand malicious activities from various attack models. It should be able to withstand and be unaffected by fake ratings or biased inputs or when there are significant patterns in the data over time. For example, an author can post fake positive ratings for his/her book while posting negative reviews for his/her competitor. Similarly, a movie can post fake positive reviews and post biased reviews against other movies. So the recommender system should be able to withstand such types of attack models while giving correct recommendations (Young and Rasik 2012; Viet-An et al. 2008).
h. **Scalability**—While the process of collection of a larger number of user ratings has become easier, this in turn has led to the growth in the size of the data sets over time. So it has now become imperative that recommender systems be designed in such a way that they can handle large volumes of data effectively and efficiently. The parameters to measure the scalability of the system are training time—i.e., the overall time to train the model should not be very high, prediction time—the time taken to generate the recommendations for the user should be low, and memory requirements—sometimes when the rating matrices are large, it is a big challenge to store the entire matrix in the main memory, then algorithms to minimize memory requirements should be designed.

9.3 Overview of the Ways to Design a Recommendation System

In the previous section, we summarized the evaluation parameters for designing an efficient recommender system. Here we give an overview of the ways for designing such a system. Although machine learning is the preferred technology for building recommender systems there are other simpler solutions also if we have fewer amounts of data or we want to quickly build a minimal solution.

Popularity-based—This is the easiest way to build a recommender system. Here the popular products are identified and recommended, e.g., in an automobile company, the most popular model is selected based on the number of cars sold. However, the limitation of the system is that it is difficult to do personalization here.

Classification-based—The second way is to use a classification model, where the features of both the users as well as those of the products are used to predict whether a product will be liked by a user or not. So for a new user, the model will collect user features like age, salary, gender, purchase history, etc., and product features like cost, quality, and product history, will be fed as input in the classifier to get a binary output of yes or no (Fig. 9.2). Additionally, this model can also capture the context like time of day or location etc. It also works well with limited user history. The limitation of this model is that sometimes all the features may not be available and secondly it does not perform as well as the collaborative filtering approach which is mentioned next.

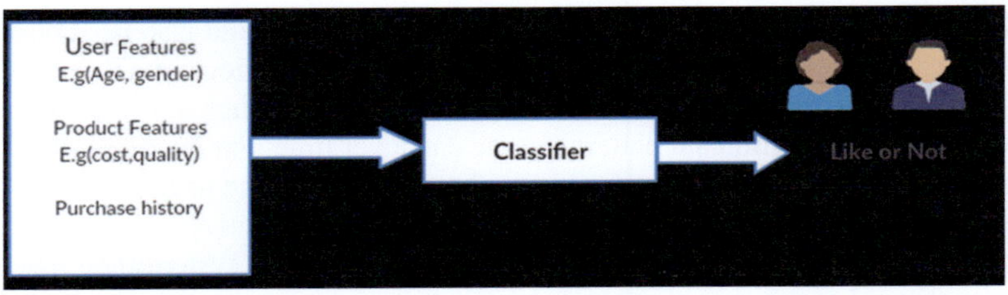

Fig. 9.2 Classification model

Collaborative Filtering

This model is based on the assumption that people like things similar to what they have already liked and also that people with similar preferences are likely to choose similar things (Guo 2013; Cai-Nicolas 2005).

There are two types of collaborative filtering models—nearest neighbor and matrix factorization.

Nearest neighbor collaborative filtering

Here the nearest neighbor (Fig. 9.3) approach is used to find similar users or similar products. There are two basic ways to filter the information for users—namely Item-based collaborative filtering and User-based collaborative filtering.

The *user-based collaborative filtering* finds users who have the same preferences and tastes in products as the current customer and have similar purchasing behavior. These choices are then recommended to the new user.

The *item-based collaborative filtering* has a different approach. It will recommend products that are similar to the product the user has already purchased, e.g., if a user

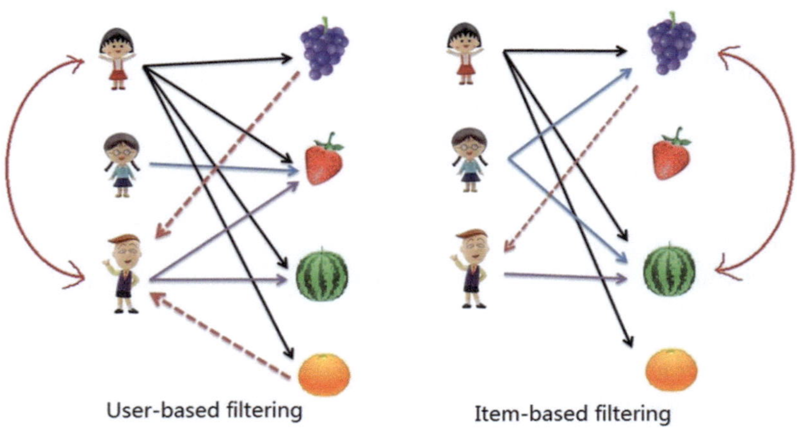

Fig. 9.3 Nearest neighbor collaborative filtering

has already liked a movie X, then a movie recommender system will try to find movies with similar characteristics and recommend those movies. The parameters to be considered for the matching could be producer name, actors, genre, release date, etc.

Matrix Factorization

The matrix factorization method is another class of collaborative filtering algorithms that are implemented in recommender systems. This method decomposes the user-item interaction matrix into the product of two lower-dimensionality rectangular matrices. This category of algorithm became very popular when it was implemented in the Netflix prize challenge and Simon Funk 2006 showed his findings to the research community to show highly effective it was. The results of the prediction can be improved if different regularization weights are assigned to the latent factors based on the item popularity and user activeness.

In this method, when a user gives feedback about any particular product, the feedback is collected and stored in the form of a 2×2 matrix. The rows of the matrix represent the different users and the columns of the matrix represent the different products. The resulting matrix is mostly sparse because not all persons will buy every product (Fig. 9.4).

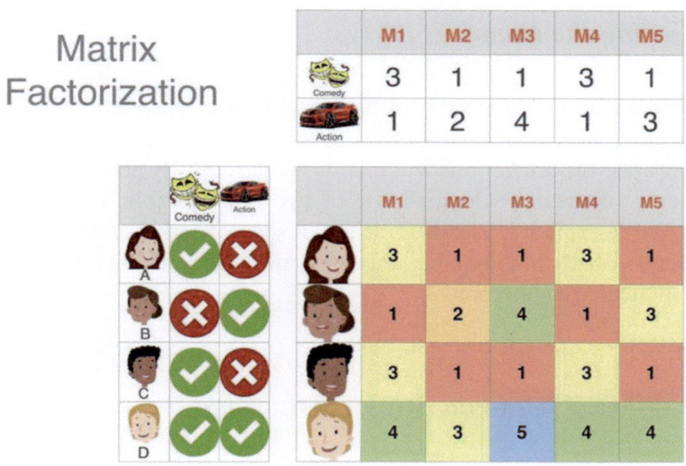

Fig. 9.4 Matrix factorization (https://www.youtube.com/watch?v=ZspR5PZemcs)

9.4 Steps in Building a Successful Recommendation Engine

To build a successful recommendation, there needs to various steps to be considered while converting a raw data to a prediction (Mano et al. 2005; Chein-Shung and Yu-Pin 2007). Normally there are six fundamental steps which we will discuss:

9.4.1 Understanding the Business

This is the first step while designing a successful system. We need to identify what are the goals of the system and what is the type of business and its special or typical features. For this there needs to be inputs from the operations team, the products team and the advertising team. The points that need to be discussed are: what the end goal of the business is, whether it will benefit from recommendations, at which point ill the recommendations occur, the availability of the data on which the recommendations will be based, whether all the contents or products should be equally treated and whether we can segment users with similar tastes.

9.4.2 Getting the Data

Since recommendation systems rely on data to make accurate predictions, the amount of data for most successful systems runs to terabytes. So the more data they have, the more accurate the predictions will be. This data is mostly about user preferences and is based on feedback which can be of two types—explicit and implicit user feedbacks. While explicit feedback is clear as it states the likes and dislikes of a user, implicit user feedback is something a user has not mentioned in his/her profile and is more complicated to interpret.

9.4.3 Explore, Clean and Augment the Data

One needs to consider the changing tastes of users while exploring and cleaning the data. So to keep up with a user's current tastes, older data that may not be relevant any more should be eliminated periodically or add a weight factor to more recent activities. It is challenging to work with datasets for recommendation systems because they are usually higher-dimensional data and a lot of the times they do not have any values, so clustering and outlier detection is difficult.

9.4.4 *Predict the Ranking*

Based on the previous steps, one can build a recommendation engine by just ranking the scores of users and get product recommendations. So there is no need to apply machine learning here and can be used for some simpler types of use cases. But for more complex ones, there needs to be more sub-tasks to be done for the further refinement of the system. This can be done by either combining the recommendations from different types of systems, use of multiple algorithms in parallel, or using pure machine learning approaches for combining multiple recommendation systems.

9.4.5 *Visualizing the Data*

Data visualization in recommendation systems serves two main purposes: firstly, it can help in revealing things in the data which would otherwise have been difficult to find and secondly, it helps in finding important information like whether some content does well but isn't being discovered, the similarities between users' tastes, which content or products are commonly consumed together, etc. so that they can make the necessary changes.

9.4.6 *Iterate and Deploy the Models*

In the final stage there is deployment of the model, and including a feedback into the loop, so that a number of iterations can be run to improve and refine the model.

9.5 Architecture

Usually, most recommender systems have three major components, an input, a recommendation algorithm, and an output. The general structure of a recommender system is shown in Fig. 9.5. The input consists of two steps. At first, we find out what is the history of the user's interaction concerning various items. The representation of this data depends on the recommendation algorithm being used and can be a vector, a matrix, or a tensor. In the majority of the recommendation algorithms, the input is represented as a matrix of ratings. It is an $m \times n$ table, where the rows represent the users and the columns represent the various products, and the intersection of the rows and columns represents the rating given by a user for that particular item in the column. If the slot is empty, it means that particular product has not yet been rated by that particular user.

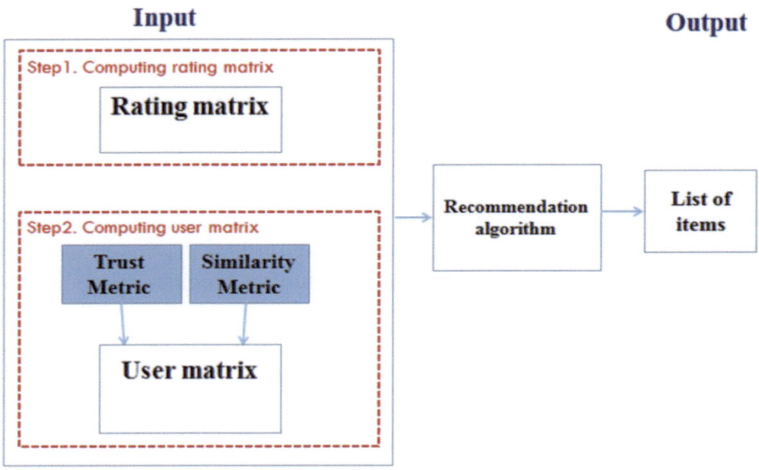

Fig. 9.5 General structure of a recommender system

The second step involves the calculation of the distance between the target users and the other users for finding their nearest neighbors. This results in an $m \times m$ matrix, where m is the number of users and the contents of any cell, i, j, is the trust entity between users u and v. While in the traditional approaches the neighbors are chosen based on similarity, now most of them are chosen based on the trust entity.

After that a suitable recommendation algorithm is applied, where the objective is to find the missing entries in the rating matrix.

Finally, the output contains the list of recommendations of products that are predicted to be of the highest preference to the user.

9.6 Summary

Thus in this chapter, we have given an overview of the steps to be followed while developing a new recommender system from scratch. We have also reviewed the parameters for the evaluation of an efficient recommender system, So depending on the type of application for the recommendation, it may sometimes be necessary to shuffle the order of priorities of the parameters to get a balanced, accurate, and relevant output.

A brief discussion of the most commonly used recommendation algorithms is also given here so that a new developer can decide the best method for his/her application. Finally, the general architecture of a recommender system has been given, and it may be customized to add more details based on the specific requirements of the application.

Think Tank

1. *What are the steps in building a recommendation engine?*
2. *What are the parameters for measuring the efficiency of a recommendation system?*
3. *What is the general structure of a recommendation system?*
4. *What are serendipity and scalability?*

References

Breese JS, Heckerman D, Kadie C (1998) Empirical analysis of predictive algorithms for collaborative filtering. In: Proceedings of the fourteenth conference on uncertainty in artificial intelligence, pp 43–52

Cai-Nicolas Z (2005) Towards decentralized recommender systems. PhD thesis, University of Freiburg

Chein-Shung H, Yu-Pin C (2007) Using trust in collaborative filtering recommendation. In: New trends in applied artificial intelligence

Gediminas A, Alexander T (2005) Toward the next generation of recommender systems: a survey of the state of the art and possible extensions. IEEE Trans Knowl Data Eng 17:734–749

Guibing G, Jie Z, Neil YS (2013) A novel Bayesian similarity measure for recommender systems. In: Proceedings of the 23rd international joint conference on artificial intelligence (IJCAI), pp 2619–2625

Guo G (2013) Integrating trust and similarity to ameliorate the data sparsity and cold start for recommender systems. In: Proceedings of the 7th ACM conference on recommender systems (RecSys)

Jennifer AG (2005) Computing and applying trust in web-base social networks. Thesis

Mano P, Dimitris P, Themistoklis K (2005) Alleviating the sparsity problem of collaborative filtering using trust inferences. In: Trust management

Paolo M, Paolo A (2007) Trust-aware recommender systems. In Proceedings of the 2007 ACM conference on recommender systems, pp 17–24

Roger C, James HD, Schoorman FD (1995) An integrative model of organizational trust. Acad Manag Rev 709–734

Viet-An N, Ee-Peng L, Jing J, Aixin S (2008) To trust or not to trust? Predicting online trusts using trust antecedent framework

Young AK, Rasik P (2012) A trust prediction framework in rating-based experience sharing social networks without a web of trust. Inf Sci 191:128–145

Chapter 10
Recommender System for Health Care

Abstract As the recommender system is applied in more and more areas, the healthcare RS is produced and has come into use for decades. In the healthcare or medical area, advice or suggestions are given for different topics like diagnosis, medicine, food, and exercise. While healthcare recommender system (HRS) can do lots of medical and fitness suggestions work accurately, several defects and optimizable functions exist in the current stages of the recommender system. To handle these drawbacks, there are several methods and techniques to be applied.

Keywords Healthcare recommender systems · Diagnosis · Medicine · Lifestyle · Food recommendation

10.1 Introduction

In recent decades, with the improvement of people's living standards and the rapid development of medical treatment levels, the medical information available on the Internet has increased significantly. In such a trend, more and more people demand better medical care or health care. To meet the popular requirements and avoid issues caused by information overload, healthcare search engines and recommendation systems are invented. Search engines filter and retrieve information through directions known to the user, while recommender systems generate information through directions unknown to the user. To some extent, these two methods can cope with most of the user's requirements and issues.

Healthcare recommender systems have been widely applied in many areas of medical treatment and support like diagnosis, medicine, and lifestyle (food, exercise, daily routine) recommendations. HRS includes several aspects of people's requirements at different levels such as therapy decisions and food and medicine suggestions. After surveying and analyzing several different kinds of recommender systems, although those available on the market can meet their target users' requirements well and perform effectively and precisely, some challenges and gaps still

P. Kar et al., *Recommender Systems: Algorithms and their Applications*, Transactions on Computer Systems and Networks, https://doi.org/10.1007/978-981-97-0538-2_10

exist between their expected performance and current conditions. Briefly speaking, these issues are the main content of systems' further improvement and updates.

In many people's thoughts, the ideal healthcare recommender system can handle any emergency or daily healthcare-related cases and meet every requirement perfectly. But in most current existing healthcare recommender systems, they can only give recommendations for a specified purpose like recommending medicine, food, exercise, or giving diagnosis and related therapies. Modern medical or healthcare suggesting procedures include several steps. Take disease diagnosis and treatment as an example, the first step is to give a diagnosis and generate accurate therapy. This therapy includes several aspects like curative medicine, follow-up lifestyle adjustment, and post-cure diagnosis. Apart from the accuracy issues and the traditional recommender system problems, the connection among several different categories of recommender systems can affect the user experience significantly. In this example, different recommender systems work independently without effective data communication which means that some parts of the recommendation may be too generic and cannot consider specific patient's conditions. Also, the work efficiency and accuracy would be rather unreliable which means that more work will be done manually by healthcare workers or common users. Beyond these connection problems, the cold start, data sparsity, and other accuracy challenges should be positioned and solved.

This document mainly mentions the challenges that exist in the system and how can developers improve their system's performance and user experience. In the first part, the document describes and analyzes the current HRS and the technologies/approaches behind it. The second focuses on issues in current HRS and suggestions on how to handle them.

10.2 Analysis of Healthcare Recommender System

The recommender system is invented to handle information overload situation in the healthcare field, and this issue exists due to the improvement of medical standards and the rapid development of the Internet. In order to give users effective and precise medical support, it is rather important to make use of a large amount of medical data efficiently and specialized healthcare recommender systems are applied to handle these data. In general, it can help doctors discover the vast amount of medical knowledge hidden in historical medical records and solve the problem of prescription information overload. For patients and other users, it can give them a better user experience and more effective healthcare support and service. In Fig. 10.1, the ideal cooperative works done by different aspects of recommender systems can work as shown in this diagram.

In the healthcare field, there are plenty of different aspects and so do its recommender system categories. In different aspects of the recommender system, the key points of the recommender system are different. For example, HRS for diagnosis has several topics like therapy decision (Gräßer et al. 2017), personalized clinical

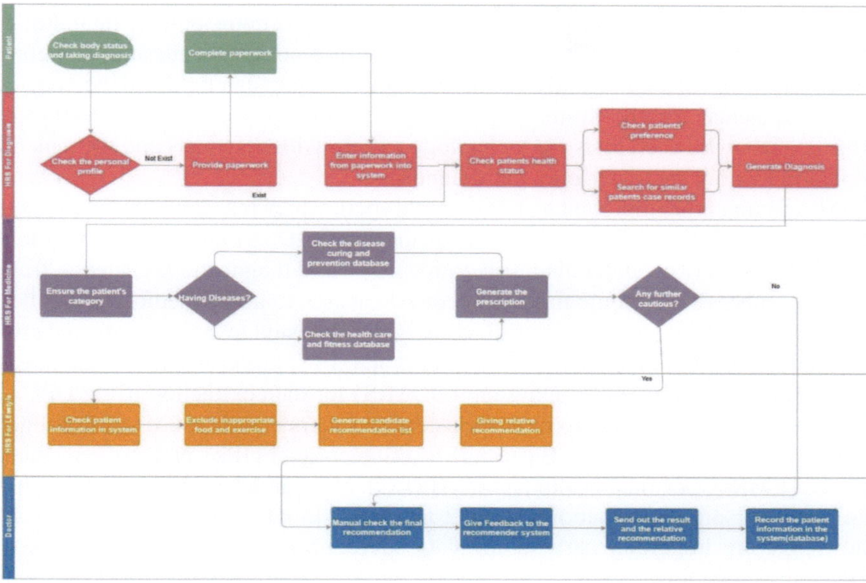

Fig. 10.1 Basic procedures for an HRS

prescription (Zhang et al. 2015), symptom analysis, and diagnosis (Gräßer et al. 2017). Also, a cooperative relationship exists between them like when giving therapy toward a specific patient case, the whole recommendation procedure involves diagnosis recommendation, medicine recommendation, and lifestyle recommendation, while therapy has different aspects of suggestions and prescriptions.

Medical recommendation systems often provide users with information such as dietary recommendations, exercise advice, and recommendations for medications and treatments (Zhang et al. 2015). In addition to the public, there are also recommendation systems for medical professionals (Tran et al. 2021), physicians can also make better decisions with the help of this recommendation system (Stark et al. 2019).

10.2.1 HRS for Diagnosis

10.2.1.1 Overview

The maturity of online access technology and the gradual improvement of medical information have facilitated the development of recommendation systems for diagnosis and treatment. In areas where medical care is less developed, doctors often do not have enough experience and knowledge to deal with various diseases, and a system like this can help local patients get better diagnostic resources. In addition, different departments in hospitals sometimes do not communicate and cooperate effectively with each other, and a recommendation system can fix this deficiency.

The audience for the system's programs is two types of people: healthcare professionals who need assistive technology to help with diagnosis and the general population who want to perform self-diagnosis.

Data mining is an important technique in this part, which is a technique to discover and extract the hidden pattern information in the dataset. Data mining has been widely used in the field of medical diagnosis to identify valid diagnostic knowledge from a large number of electronic medical record databases to assist medical personnel. Commonly used data mining algorithms are KNN, decision tree algorithm, and artificial neural network algorithm. Medvedeva et al. (2017) proposed disease diagnosis and treatment recommendation system (DDTRS) for data extraction using DPCA, and they first group reports to obtain clustering centers and then use Apriori algorithm for association analysis. Electronic health records (EHRs), electronic medical records (EMRs), personal health records (PHRs), etc., can be used to mine patients' cases and personal information (Stark et al. 2019). In addition, expert advice and knowledge are important parts of the data source. Some systems use sensors to collect key physiological data from patients, such as blood pressure, heart rate, and other information.

Taking judging the probability of cardiovascular disease and its related symptoms as an example, the recommender system will build a probability table by applying the Bayes network methodology. Figure 10.2 shows the relationship between the symptoms and heart diseases and gives the calculation procedures of the operating principles of Bayes network for HRS.

Patient similarity analysis is also an essential part of the process. Using collaborative filtering techniques to find similarities between patients based on symptoms has good predictive power. Equation (10.1) shows similar calculation procedures when giving a diagnosis to the target patient by applying Pearson correlation coefficient (Jain et al. 2020). The previous patient's case with the highest similarity score will be considered the most suitable reference when giving a diagnosis.

$$\mathrm{Sim}(a, b) = \frac{\sum_{p \in P} (r_a - \overline{r_a})(r_b - \overline{r_b})}{\sqrt{\sum_{p \in P} (r_a - \overline{r_a})^2} \sqrt{\sum_{p \in P} (r_b - \overline{r_b})^2}}, \tag{10.1}$$

while

P	means the attributes' set that contains the useful diagnosis data.
Patient a	is the target patient and
Patient b	is the recorded patient case.
r_a and r_b	are attribute p's data from Patient a and Patient b.

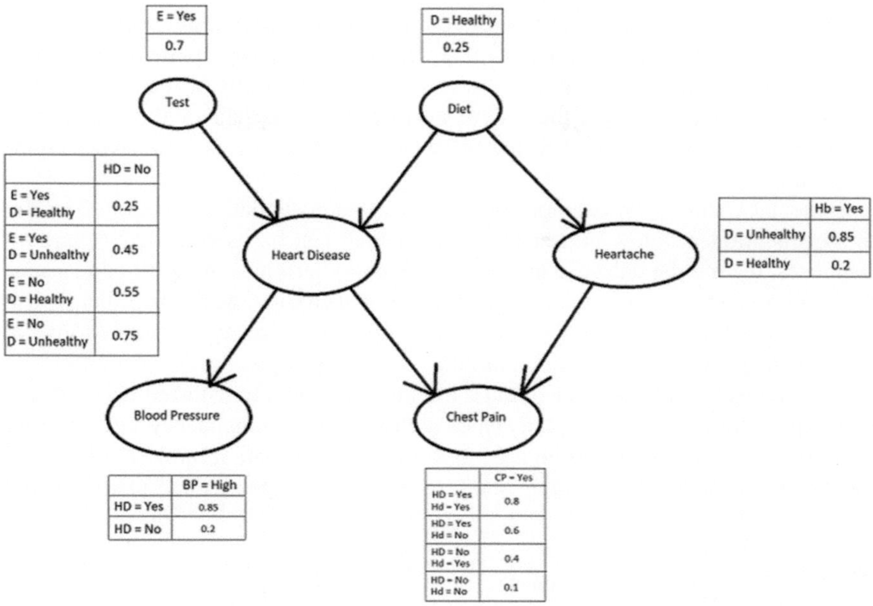

Fig. 10.2 Probability table of heart diseases diagnosis (Bayes network)

In addition to similarity calculation based on symptoms, metrics such as consultation fees and ratings of doctors can also be used to find patients with the same preferences as the target user. Collaborative recommendations and demographic-based recommendations were used in this category of recommender system.

Research is conducted for diabetes diagnosis systems, using a collaborative filtering framework, and using both clustering and classification methods for a recommendation, which effectively reduces the complexity of high-dimensional data and improves the computation time. However, it has also been noted that classification-based approaches are superior to collaborative filtering approaches.

10.2.1.2 Case Analysis—Therapy Decision

When making therapy decision for patients, the preciseness and accuracy must be guaranteed, or it will cause misdiagnose and leads to serious consequence.

In this case, the recommender system should collect several data like basic information about the patients (age, gender, BMI, family anamnesis), therapy description information (a medicine used in this therapy, curative period, dietary consideration), past patients' records used to do a comparison (this part is for the therapies which have been widely used in curing patients), clinical trial data (this part is for the therapies which have not been used and can be considered as new-produced therapy). These data are used commonly when making therapy decisions because most diseases can be estimated through these indexes. In addition to these data, more specified information needs are to be collected. For example, when estimating a patient, the severity of his/her Psoriasis (a kind of skin disease) (Gräßer et al. 2017), the system needs to collect the health status of the patient's skin while these data are the main factors to judge. Also, the system will not collect other irrelevant data to save memory resources.

Table 10.1 shows the example of Psoriasis diagnosis data collection. The first dataset mentioned below is user attributes, which can be used in many diseases' diagnoses, the middle dataset can be changed into other kinds of specified measurement indexes according to the actual application scenarios, while the other dataset is designated for Psoriasis only like PSAI scores and severity in order to give an accurate diagnosis to this specified disease.

After collecting all the needed data, the information is transferred into different categories of value that can be easily calculated by the recommender system program. All the categories of data can be used as components of data mining. For example, the above table's data has some general information like gender and weight, which

Table 10.1 Example for Psoriasis patient describing attributes (Gräßer et al. 2017)

Attribute	Scale	Range
User attributes		
Year of birth	Interval	1931–1998
Gender	Nominal	1, 2
Weight	Ratio	50–165
Size	Ratio	99–204
Family status	Dichotomous	0, 1
Planned child	Nominal	1, 2, 3
Year of first diagnosis	Interval	1950–2014
Dataset according to the actual application scenario		
Type of psoriasis	Nominal	1, 2, 3, 4, 5, 6
Family anamnesis	Ordinal	1, 2, 3
Comorbidity	Nominal	1, 2, 3, …, 34
Status	Ordinal	1, 2, 3
Under treatment	Dichotomous	0, 1
Dataset for psoriasis only		
Disease-free	Dichotomous	0, 1
PASI score	Ratio	0–43
Severity	Ordinal	1, 2, 3, 4, 5
Development face	Ordinal	1, 2, 3 (severity index)
Development feet	Ordinal	1, 2, 3 (severity index)
Development nails	Ordinal	1, 2, 3 (severity index)
Development hands	Ordinal	1, 2, 3 (severity index)

may have potential effects on the final diagnosis. Also, it contains some direct data like the development part of body skin and they play an important role in generating the therapy choices list.

This HRS' drawbacks are obvious that all the therapy recommendations should know what the patient's disease is in advance or the RS cannot make any accurate decision. Before applying this recommender system, the disease diagnosis (symptoms analysis and diagnose) RS should be applied and recognize the disease as accurately as possible. Also, the most used RS in the current stage is classification-based, which may give overbroad therapies a selective list and lead to poor curative efficiency and low accuracy. To handle this issue, traditional data mining techniques should be improved, and the restriction of descriptive data should be overcome.

Besides, the traditional issues of RS also occur in this category of HRS. The sparsity of the data is one of the most important problems of this system, which can be compensated by a demographic-based technical approach, as the latter has broader coverage. One reason why this problem exists is that patients hold back information about themselves due to privacy concerns. Another reason is that doctors do not require all physical tests when diagnosing a patient. As for the cold start problem, it can be effectively mitigated by importing many users' prior cases. The merging of home healthcare data is also a difficult problem, which involves the limitation of high-dimensional data.

10.2.1.3 Challenges and Solutions in HRS for Diagnosis

When giving a diagnosis to patients, the most important target is to make no mistakes. In existing recommender systems, the most applied measurement methods depend on keyword analysis instead of the quantitative index while the recommender system can handle figures better than keywords. For example, in Cardiovascular Risk Computation Recommender System (Guzmán et al. 2018), the programmers developed a decision tree to judge whether the patient has cardiovascular diseases. The decision tree shown in Fig. 10.3 has several keywords like heart frequency, lifestyle, cardiovascular risk, and current exercises, and under these nodes, there is 2–4 keyword such as sedentary, normal, and active for describing the patient's lifestyle. Compare with other quantitative indexes like blood pressure and heart rate, these keywords are too broad for doctors to make a judgment. This method may be rather good for data mining when there exists a large enough dataset for a recommender system to train and judge. If the number of cases is small, then the system will be more likely to give inaccurate judgments.

To handle these issues, the main point is to use more quantifiable factors like heart rate and several categories of blood indexes which can be managed intuitively by a recommender system rather than vague keywords. To some extent, the recommendation will mainly focus on the quantizable rather than simply Boolean operators or choices. Besides, for some rare disease which has little or even no data stored in the background database, the recommender system cannot give valuable suggestions

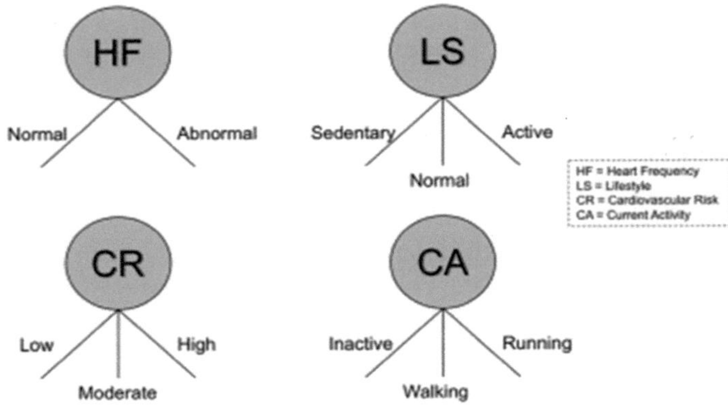

Fig. 10.3 Decision tree for cardiovascular risk computation

Table 10.2 Table of the patients' body index

Case ID	Age	BMI	Insulin	BP	Glucose
1	25	27.9	25.4	67	86
2	27	25.8	25.9	72	166
3	48	46	33	82	118
4	27	31	28	70	89
5	53	31.7	43.2	69.7	195

and these cases should be handled by doctors themselves. Table 10.2 gives some commonly used body indexes that can measure users' health status.

All the patients' cases (clinical information) are stored in the background database and each risk factor is assigned a score to represent its impact degree on the patient. Also, the patient's information can have other kinds of data but are not shown in this table due to the irrelevance of data and they may not have any effects to estimate patients' health status. Besides the information given above, all the cases will be clearly labeled whether the patients have diabetes or not. After that, these data can be treated as training sets and entered into the recommender system to make it give a more accurate recommendation (diagnosis).

10.2.2 HRS for Medicine

10.2.2.1 Overview

With the rapid development of medical technology and pharmaceutical level, more and more medicines are invented and applied in different areas like specific medicine of disease and medicine to prevent specified diseases. While the variety of medicines has increased significantly, the probability of accidents has also increased. According to the FDA report, more than 42% of medication errors are caused by physicians. The human brain can sometimes make mistakes. A drug recommendation system can be a good help to solve this problem. However, drug recommendation systems require access to very specialized medical information. Data specialization in the drug field may be more stringent than in other fields. A movie recommendation list can occasionally contain content that is not of interest to the user, but a drug recommendation list should never contain drugs that cause illness or death. The machine must recommend the right drug and needs to identify multiple correct drug interactions and adverse reactions to related drugs. Some ontology and rule-based drug recommendation systems recommend patients by analyzing detailed information about the drug itself. González et al. (2009) used an ontological approach to diagnose specific diseases. It uses only three variables: type of disease, drug, and allergy rules.

To some extent, HRS for medicine can be used by working cooperatively with HRS to diagnose when giving therapy to patients. For the commonly used medicine, the HRS background database will clearly label it with its attributes like target disease, curative period, and market price. And for the new-produced medicine, the HRS will analyze its clinical data and ingredients to extract useful information and then store the medicine in the background database. When searching for the most suitable medicine for patients with a specific disease, the HRS will compare the patients' profiles and medicine database and generate a selective list for the attending doctor to decide.

As the level of modern pharmaceuticals improves significantly, there are so many kinds of drugs for different purposes on the market today which means that the HRS for medicine should give appropriate recommendations according to the medical or healthcare purposes. To handle this challenge, the medical records stored in the background database are clearly labeled with medical purpose, usage cautious, potential, or direct side effects and other necessary attributes which may influence the final recommendation. Table 10.3 shows the background dataset structure, Table 10.4 shows the relative recommendation procedures used to cope with the challenge, and Table 10.5 gives a simple description of these three categories. The medicine database can be mainly divided into three different types according to the user group.

Also, prediction of drug response is an important research direction, and prediction of drug side effects and interactions or prediction of patient response to specific drugs might take the drug recommendation system to a new level. A combined sparse linear recommendation and logistic regression model (SlimLogR) was used in Chiang et al. (2018) to solve the recommendation problem, which contains a drug recommendation component and a label prediction component that efficiently identifies decisions that may cause adverse drug reactions. The effects of drugs for some cancers or rare diseases may vary from person to person, and in the era of precision medicine, one can predict a patient's response to a specific drug by analyzing data on the patient's gene expression, rather than relying solely on the large number of cases that existed in the past to make predictions. Suphavilai et al. (2018) combining different cell lines

Table 10.3 Medicine database structure and description

Key name	Type	Range
Database structure (data stored in the database)		
Medicine ID	TEXT	Unique information, primary key
Medical purpose	TEXT	Fitness ǀ cure ǀ prevention
Side effects	TEXT	(According to details in practice)
Ingredients	TEXT	(According to details in practice)
Authority level	INTEGER	1 ǀ 2 ǀ 3 (user group index)
Instruction	TEXT	(According to details in practice)
Postscript	TEXT	(According to details in practice)

Table 10.4 Assessment levels of a different user group

Common patients/users	Common healthcare workers	Professional healthcare workers
Access level of the different user groups		
Medicine not prescribed by a doctor Low risk and minimal side effects Mostly for daily healthcare and disease prevention	Most medicine can be recommended except for some risky medicine Can handle most of the medical cases Need manual check before giving a recommendation	All the medicine can be recommended, and more detailed information will be provided Have authority to modify the information in the medicine database Need manual check before giving a recommendation, and this step cannot be ignored

Table 10.5 Medicine database structure and related procedures

Common patients/users	Common healthcare workers	Professional healthcare workers
Database searching procedures		
Check user's requirements for keeping fit Digitalize user's health status and enter them into the system Narrow the selection range and generate a recommendation list	Collect information about the target patient and his/her diseases and symptoms Digitalize that information and enter them into a system Generate a recommendation list and ask workers to check and estimate twice Give patient prescription	Collect information about the target patient and his/her diseases and symptoms Digitalize that information and enter them into a system Generate a recommendation list and access feasibility and risk Give patient prescription and detailed cautious

with specific drugs and exploring their potential relationships lead to understanding drug mechanisms and recommending accurate drug regimens more effectively. Drug side effects are a potential factor in public safety and one of the major contributors to illness and death in health care (Tatonetti et al. 2012). Galeano and Paccanaro (2018) it proposed a collaborative filtering model for the large-scale prediction of drug side effects that can be used for the early detection of adverse drug events.

10.2.2.2 Case Analyze—Medicine Side-Effect Prediction

Medicine side effect is an important criterion when giving recommendations and cannot be ignored in most cases. The American Institute of Medicine reported that 100,000 deaths occur annually in the USA from medical errors, many of which are caused by unexpected drug side effects (Yera Toledo et al. 2019). When selecting medicine for therapies, the HRS should attach great importance to predicting new-produced medicine's side effects and analyzing recorded medicine's side effects.

Many drugs have side effects. They can be categorized by their degree of influence from mild to severe and linked with their cured diseases. When the drug's side effect is mild and acceptable to the patients, the system should take it into account as the other medicine with no side effect but are slightly different. In the recommendation system, we can attach the weight calculated by weighing pros (curative effects, curing period, market pricing) and cons (side effects, directly or latent) to the given drugs to give the best recommendation with the highest curative effects and lowest side effects.

In the existing mature recommendation systems of side-effect prediction, the most commonly applied techniques are the latent factor model (Galeano and Paccanaro 2018) and artificial neural network (ANN). The following model shows a neural network of how the side effects in the bottom layer are generated from the drug. Top layer nodes $p_1, \ldots, p_k$ provide a hidden representation of a drug, while bottom layer nodes $r_1, \ldots, r_n$ contain the scores for all the potential side effects for the given drug. Figure 10.4 shows the ANN structure of side effects' prediction.

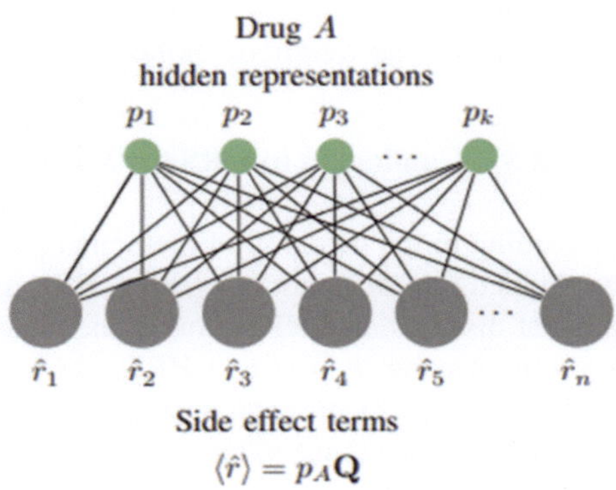

Fig. 10.4 ANN for side effects' prediction

This method is based on an authoritative medicine database that stored all the information about medicine attributes like target disease, ingredients, and latent side effects. For example, the side-effect resource (SIDER) version 4.1 is one of the mature healthcare databases which contains information about medicines on the market and their recorded side effects (Kuhn et al. 2015).

Besides, the similarity measurement methodology can be used in this category of recommendation such as the Jaccard similarity index calculation. To predict the drug's side effects, the first step is to find the group of drugs with similar curative effects' ingredients and other attributes. These features should be indexation and given related weight during the calculation step. The similarity equation can be mentioned as the weight of the features that two or more drugs both have to be divided by the total weight of the features that all the drugs in this group have. The similarity equation and estimation equation are given in Eqs. 10.2 and 10.3, respectively.

$$\mathrm{Sim}_{\mathrm{drugs}} = \frac{\sum_{i=1}^{n} w_n \prod_{i=1}^{m} a_{m,n}}{\sum_{i=1}^{n} w_n}, \tag{10.2}$$

while

w_n means the weight of a kind of side effect *n*.
$a_{m,n}$ means whether the drug *m* has the side effect *n*.

After collecting the dataset, the second step is to predict the target drug (new-produced drug) side effects. The calculation result will show the side effect with the highest probability, and it will be considered as the potential side effect of the target drug. The equation can be written as:

$$\mathrm{Prob}_{\mathrm{drug}} = \frac{\sum_{i=1}^{n,m} \mathrm{Sim}_m w_n}{\sum_{i=1}^{n} w_n}, \tag{10.3}$$

while

Sim_m means the similarity index between the target drug and the searching drug *m*.
w_n means the weight of a kind of side effect *n*.

10.2.2.3 Challenges in HRS for Medicine and Possible Solutions

According to the diversity of medicinal purposes, the recommender system should be divided into different categories, HRS aims at recommending medicine used to prevent diseases, and HRS aims at the medicine that can be used to stay healthy or keep fit. Among the existing HRSs for Medicine, all of them are mainly used by doctors, while patients cannot use this service due to their professionality and safety. It means that many users' requirements cannot be satisfied like choosing healthcare medicine or keeping preventive medicine on hand.

When developing this kind of HRS, the system should not only focus on the users' profiles to collect their health status but also take their preferences and further requirements into account. For example, if the target user is a normal person and wants to keep fit, the recommender system should give a healthcare medicine list for users to choose from rather than compulsory medical suggestions. Figure 10.5 is the flowchart that shows how the recommender system copes with the user's requirements with different identities and authorities.

To solve this issue, the recommender system will be mainly used by healthcare professional workers if the recommended medicine is used for curing or preventing diseases, so they will know the purpose why the target medicine should be used. When giving personalized medicine prescriptions, the objective purpose should be confirmed, and it needs the background medical database to store the purpose of different medicine. Also, when the objective purpose of medicine is used to keep fit or for daily health care, the main consideration for generating the recommendation is the user's preference. Based on this idea, the Naïve Bayes classifier can be applied and optimized to the recommendation list.

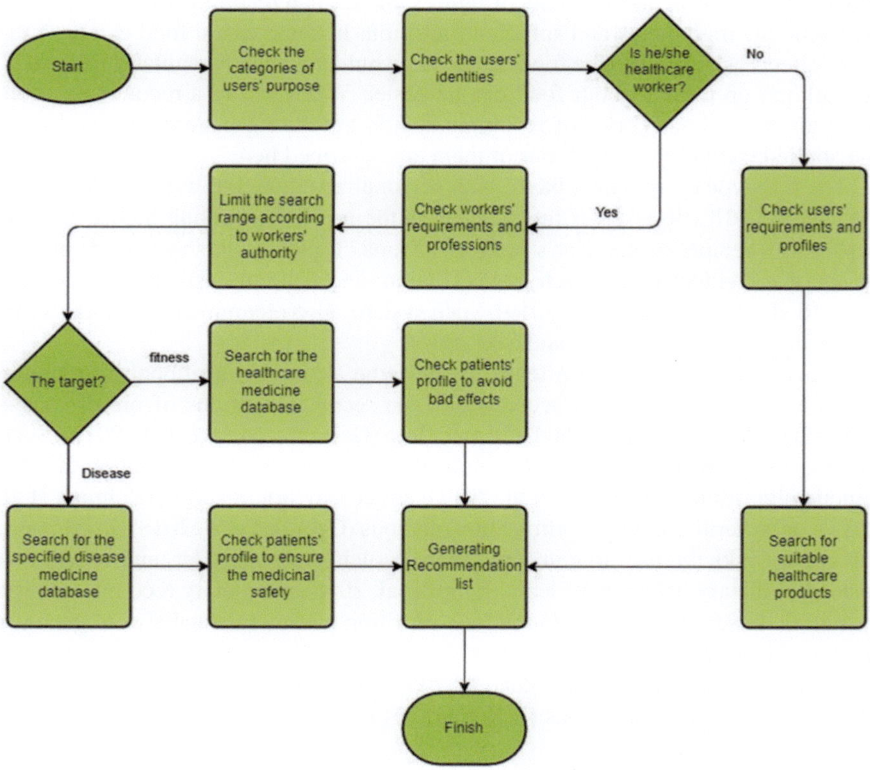

Fig. 10.5 Flowchart of improved medicine recommendation

According to the attributes of different medicines, the background database should store the medicine information like known side effects, critical ingredients, taking instructions, and curative effects. All this information will be recorded in a standardized format to be efficiently handled by the recommender system.

10.2.3 HRS for Lifestyle

10.2.3.1 Overview

HRS for lifestyle can be divided into two categories, HRS for food and HRS for exercise. The food recommendation system or diet recommendation system performs like the drug RS to a certain extent. When recommending food for certain diseases, the recommendation system should take the user's lifestyle and health conditions into account. This part of the recommendation is mainly based on the user's dietary habits and current health conditions like BMI, blood pressure, blood glucose, and blood fat. According to these data or figures, the RS can recommend specified food with suitable ingredients and sufficient nutrition. In some cases, food can have the same effects as drugs do. The food categories would not change regularly like drugs, so building up the RS about food can be easier. When giving a recommendation, there are several categories of recommendation-generating patterns depending on the applied technologies, and four of them are described here.

The first type of system is based on a personalized model of users (Galeano and Paccanaro 2018). It asks for inputs such as the user's past eating habits, such as yesterday's recipe, or the user's preferred foods. Typically, this system displays a series of individual foods, such as beef, lamb, and fish, and asks the user to rate these foods (Yera Toledo et al. 2019). Such systems also recommend restaurants that match the user's tastes (Tung and Soo 2004).

The second type of system tends to study the available nutritional information and process the information according to the recommendations of professionals, rather than prioritizing individualized modeling (Galeano and Paccanaro 2018). Such systems use existing healthcare recommendations and process information through genetic algorithms (Syahputra et al. 2017), ant colony optimization (Rehman et al. 2017), or bacterial foraging optimization methods (Chouhan et al. 2018). In this case, the system asks the user to enter information such as age, gender, and occupation instead of dietary habits. NutElcare (Espín et al. 2016) is a dietary recommendation system that takes recommendations from nutrition or dietetic databases to generate reliable dietary recommendations for older adults. GPS and pedometers are used to estimate how much nutritional intake the user needs for the day and then generate a ranked list of recommendations (Nag et al. 2017).

The third meal recommendation system blends the first two approaches, which effectively combines user preferences and nutritional recommendations. Trattner and Elsweiler (2017) presented a pioneering approach that investigated the possibility of including nutritional factors in recipes. Ge et al. (2015) and McCarthy et al. (2016) have made great efforts in this regard, being able to balance the user's taste and healthy dietary requirements.

The fourth type of recommendation system is a group of people who recommend a common diet (Galeano and Paccanaro 2018). For example, when a group of users is planning to have a party, this kind of system can consider everyone's eating habits to make recommendations instead of considering only one user (Kuhn et al. 2015).

Another category of this HRS is physical exercise recommendation. Compared with other kinds of healthcare recommender systems, its significance is much lower than other kinds of HRS. When recommending the exercise, the RS will take the user's physical conditions into account. Unlike the medicine and food recommendation system, the enforceability of the physical exercise recommender system is much lower while the recommendation will mainly depend on the user's personal willingness. The only factor that needs to be concerned is the user's health conditions. For example, users with leg injuries should not be recommended for long-distance races and other kinds of exercises with legs. And users with cardiopathy or asthma should not engage in vigorous exercises. To cope with this requirement, build up a don't list to record all the chronic diseases that need special attention. This information should be stored in the database.

10.2.3.2 Case Analysis—Healthcare Food Recommendation

With people's pursuit of a healthy diet, the HRS for healthy food and dietary management rose in response to people's growing requirement for physical health. As mentioned before, this kind of recommender system lacks enforcement, and people can choose their dietary plans and decide whether to follow them or not. Also, the variety of food recipes and ingredients has increased significantly. Helping people choose food properly and wisely is important in today's world.

The HRS for healthcare food recommendation works slightly differently than other HRS. It firstly collects users' basic information like preferences to reduce the probability that users do not satisfy the dietary plan and refuse to follow it. Then, the system will take the users' health status to choose which food or ingredients to recommend. For example, recommend low-glycemic foods or sugar substitutes for diabetics and food with less oil and salt for high blood pressure users. This step will generate a recommendation list of detailed dietary plans which contains guidance like eating what, when to eat, and how to eat to get better effects. After that, the recommender system will ask users to evaluate the dietary plans by scoring so that for other users, the system can compare users' profiles with other users' records collected before and save lots of execution time to generate a recommendation list.

Fig. 10.6 Cluster diagram of target people positioning

10.2.3.3 Challenges in HRS for Lifestyle and Possible Solutions

It is obvious that in this kind of recommender system, the user's preference and requirements have a higher priority and the system is of low coerciveness. But sometimes, the user's health status should be considered, and the recommender lifestyle may not be accepted by the users due to their personalities and life experience. To handle this issue, the target group positioning should be done when generating the recommendation result. KNN algorithm can be applied in this area. For example, the following figure consists of two categories of users' eating lifestyles (food taste, 0 is a light flavor and 1 is a heavy flavor, and dinner starting time). After all the data is inserted into the cluster diagram, Fig. 10.6 shows the final cluster diagram and it is easy to find the clusters.

Figure 10.6 contains several features, and in the actual condition, there will be more users' information and considered features. Beyond this diagram, all the points shown in the diagram contain case information like the recommendation list given to the users and their satisfaction with the plan which can be mentioned as the training set.

Before starting the recommender system, the program will collect the user's daily lifestyle information and insert the data point into the cluster diagram. Then, the system will find the user group that the target user belongs to and give the recommendation list like other members in the user group have taken. After that, the system will ask about the user's satisfaction with the result for further system training and provide more accurate suggestions.

10.3 Real Example

A real example of healthcare recommender system, Find a Doctor, is available in GitHub repository. Readers can access the source code of the project through the following link.

https://github.com/pushpendukar/Healthcare_RS.git

10.4 Summary

This article describes the application of recommender systems in health, including recommendation applications in four major areas: food, exercise, drugs, and diagnostics. In general, collaborative filtering is the more common application technique for health recommendation systems. Among these healthcare recommender systems, it is obvious to see the inner connections between them. For example, before giving medicine recommendations for patients, the HRS for diagnosis is applied and the patients' diseases must be determined with high accuracy. Also, the recommendation for food and medicine can be applied together, while food may have similar effects as medicine does. To conclude, a possible future research direction is a full domain health recommender system that can cover all the above recommendations, which has very high user potential and promise.

Think Tank

1. *What are the benefits and challenges in HRS?*
2. *How can HRS be used for a better lifestyle?*

References

Chiang W-H, Shen L, Li L, Ning X (2018) Drug recommendation toward safe polypharmacy. ArXiv abs/1803.03185

Chouhan SS, Kaul A, Singh UP, Jain S (2018) Bacterial foraging optimization based radial basis function neural network (BRBFNN) for identification and classification of plant leaf diseases: an automatic approach towards plant pathology. IEEE Access

Espín V, Hurtado MV, Noguera M (2016) Nutrition for elder care: a nutritional semantic recommender system for the elderly. Expert Syst J Knowl Eng 33:201–210

Galeano D, Paccanaro A (2018) A recommender system approach for predicting drug side effects. In: 2018 International joint conference on neural networks (IJCNN), pp 1–8

Ge M, Ricci F, Massimo D (2015) Health-aware food recommender system. In: Proceedings of the 9th ACM conference on recommender systems

González AR et al (2009) SemMed: applying semantic web to medical recommendation systems. In: 2009 First international conference on intensive applications and services, pp 47–52
Gräßer F, Beckert S, Küster D, Schmitt J, Abraham S, Malberg H, Zaunseder S (2017) Therapy decision support based on recommender system methods
Guzmán G, Torres-Ruiz M, Tambonero V,·Lytras MD, López-Ramírez B,·Quintero R,·Moreno-Ibarra M,·Alhalabi W (2018) A collaborative framework for sensing abnormal heart rate based on a recommender system: semantic recommender system for healthcare
Jain G, Mahara T, Tripathi KN (2020, Jan) A survey of similarity measures for collaborative filtering-based recommender system. https://doi.org/10.1007/978-981-15-0751-9_32
Kuhn M, Letunic I, Jensen LJ, Bork P (2015) The sider database of drugs and side effects. Nucleic Acids Res 44(D1):D1075–D1079
McCarthy K, Salamó M, Coyle L, McGinty L, Smyth B, Nixon P (2016)Group recommender systems: a critiquing based approach. In: IUI'06
Medvedeva O, Knox TG, Paul J (2017) DiaTrack: web-based application for assisted decision-making in treatment of diabetes. J Comput Sci Coll 23:154–161
Nag N, Pandey V, Jain RC (2017) Live personalized nutrition recommendation engine. In: Proceedings of the 2nd international workshop on multimedia for personal health and health care
Rehman F, Khalid O, Haq N, Khan AR, Bilal K, Madani SA (2017) Diet-right: a smart food recommendation system. KSII Trans Internet Inf Syst 11:2910–2925
Stark B, Knahl C, Aydin M, Elish KO (2019) A literature review on medicine recommender systems. Int J Adv Comput Sci Appl
Suphavilai C, Bertrand D, Nagarajan N (2018) Predicting cancer drug response using a recommender system. Bioinformatics 34:390W 3914
Syahputra MF, Felicia V, Rahmat RF, Budiarto R (2017) Scheduling diet for diabetes mellitus patients using genetic algorithm
Tatonetti NP, Ye P, Daneshjou R, Altman RB (2012) Data-driven prediction of drug effects and interactions. Sci Transl Med 4:125ra31–125ra31
Tran TNT, Felfernig A, Trattner C, Holzinger A (2021) Recommender systems in the healthcare domain: state-of-the-art and research issues. J Intell Inf Syst 57:171–201
Trattner C, Elsweiler D (2017) Food recommender systems: important contributions, challenges and future research directions. ArXiv abs/1711.02760
Tung H-W, Soo V (2004) A personalized restaurant recommender agent for mobile e-service. In: IEEE international conference on e-technology, e-commerce and e-service. EEE'04, pp 259–262
Yera Toledo R, Alzahrani AA, Martínez L (2019) A food recommender system considering nutritional information and user preferences. IEEE Access 7:96695–96711
Zhang Q, Zhang G, Lu J, Wu D (2015) A framework of hybrid recommender system for personalized clinical prescription. In: 2015 10th International conference on intelligent systems and knowledge engineering (ISKE), pp 189–195

Chapter 11
A Surveillance Framework of Suspicious Browsing Activities on the Internet Using Recommender Systems: A Case Study

Abstract Cybercrime activities are increasing all over the world at an alarming rate and pose a serious threat to a nation and its citizens. For this reason, surveillance and monitoring of online browsing activities of individuals become necessary to prevent terrorist and criminal activities, in the interests of national security. There are at present several such surveillance projects in implementation already in India as well as abroad. The cyber-surveillance tools monitor data stored on hard drives, as well as data transferred over computer networks, such as the Internet through emails, or through mobile phones in the form of calls or messages. In this chapter, we have proposed a sparse matrix-based framework that will track the browsing activities of individuals over a period of time, and if some persons are found to be surfing a chain of websites that are categorized as potentially harmful, for a period of time, then it will raise an alert to the governing authorities. The framework also uses recommender systems and browsing history tracking algorithms. We expect that this can be of assistance and can be utilized by authorized monitoring agencies of countries as a threat analysis tool.

Keywords Cybercrime · Surveillance · Browsing history tracking · Recommender systems · Sparse matrices · Threat analysis

11.1 Introduction

Computer and network surveillance is the monitoring of computer activities, or data stored or transferred over the Internet by individuals or organizations. These activities can be performed from laptops, PCs, mobile devices, or other devices used for electronic communication. Cyber-surveillance has helped to solve various crimes and resolved cases through proper investigation (http://donottrack.us/; https://www.ghostery.com/).

This monitoring of all types of interest traffic is usually done by organizations that are authorized by the government to covertly carry out these operations and this

P. Kar et al., *Recommender Systems: Algorithms and their Applications*, Transactions on Computer Systems and Networks, https://doi.org/10.1007/978-981-97-0538-2_11

information is usually not available otherwise. The government needs to do this to identify and monitor threats to society, nation, or the economy.

This is done with the help of surveillance tools, highly sophisticated equipment, and software (https://sharemenot.cs.washington.edu).

Criminal and terrorist activities are increasing at an alarming rate all over the world and these are being easily planned and executed by using the Internet facilities to their advantage. Hence, it is of utmost importance that a government is adequately equipped with the tools and the technology to track such activities and stop them from causing harm to anyone.

In this chapter, we have proposed a framework to track the browsing history of individual groups and build an alert system from the pattern of surfing history. If the number of accesses to certain sensitive or malicious sites from a particular machine or place is beyond a predefined counter value, then it will raise a flag to the concerned surveillance authorities, along with the browsing history and the location/identity of the user. The rest of the chapter is organized as follows: Sect. 11.2 contains the related work, Sect. 11.3 contains the details of web user tracking of browsing patterns and populating recommender systems, Sect. 11.4 gives an overview of the advantages of using sparse matrices for our framework, Sect. 11.5 gives the proposed framework, Sect. 11.6 is the proposed algorithm, and the conclusion is in Sect. 11.7. In the next section, we give an overview of the existing surveillance frameworks and their functionalities.

11.2 Related Work

There are at present several mass government surveillance projects in India, all of which are monitored by the various core security agencies authorized by the GOI. There are various other mass surveillance projects in place all over the world also. Here, we briefly mention three such major projects in India (http://trackingobserver.cs.washington.edu; Acar et al. 2014).

The National Intelligence Grid (NATGRID) contains integrated data in the form of a master database and is used for counter-terrorism operations. It integrates the databases from various important security agencies under the GOI. It was formed after the 2008 Mumbai attacks. The whole master database is scheduled to live by December 31, 2020. Very few people will have access to it on a case-to-case basis.

NATGRID is a counter-terrorism measure, which collects varied data from various standalone agencies and Indian government ministries. The collected data collate all kinds of information from these databases, like details of banking accounts, taxes, transactions on credit or debit cards, records of visa and immigration, travel itineraries, etc. With the help of NATGRID, security agencies can locate and extract relevant information about terror suspects by pooling data from datasets all over the country. Basically, it is aimed to help in the identification and capturing of suspected terrorists and thereby help in preventing plots of terror attacks.

The Central Monitoring System (CMS) works on the principle of telephone interception provisioning and is a centralized system. It was developed by C-DOT, which is a GOI-owned telecommunication technology development center. It monitors all the communications that take place on mobile phones, landlines, and the Internet in the country. It is capable of keeping track of the persons who have initiated or received calls on mobile or landline numbers, the time, date, and durations of the calls, as well as the location of the targets, failed calls, the call data records (CDRs) of roaming subscribers, and forwarded telephone numbers by target subscribers. It was designed primarily to strengthen security in the country.

The Center for Artificial Intelligence and Robotics (CAIR) has developed a software network named NEtwork TRaffic Analysis (NETRA) in a DRDO laboratory and is used by various counter-intelligence agencies like the Intelligence Bureau and RAW. Initially, this program is being tested by the national security agencies on small scale, but it is planned to be deployed at a pan India level soon. It has been designed to monitor Internet traffic on a real-time basis because of the increasing threats posed by criminals and terrorist groups who are using data communication among themselves. NETRA is capable of analyzing the voice traffic which passes through a variety of softwares like Skype, Google Talk, etc., and it can also detect and intercept messages which contain keywords like "attack", "bomb", "blast", or "kill" in real-time from a huge number of tweets, status updates, emails, Internet calls, blogs, forums, etc. It can also detect images that are generated on the Internet for obtaining the intended intelligence information. The system with RAW analyzes a large amount of international data which crosses through the Internet networks in India.

The methods mentioned above all deal with the Internet or network communication where they track tweets, emails, messages, and other electronic documents. But to the best of our knowledge, there is at present no framework to track the browsing patterns of users to predict and raise alerts for visiting links for malicious or potentially harmful websites. In this chapter, we have proposed a framework that will be able to trace the browsing patterns of various users using various parameters and alert authorities if any suspicious activity is suspected. The authorities can then further investigate the flagged users for possible threats. In the next section, we give an overview of how the web tracking of the browsing patterns is usually done along with an introduction to the concept of recommender systems.

11.3 Web User Tracking of Browsing Patterns and Populating Recommender Systems

App and service providers do extensive web user tracking to collect all information about their users, in order to provide them with a superior product experience. Information about the location, browsing tendencies, communication records, financial information, and general preferences regarding users' online and offline activities can give significant insights into a person's activities. A lot of this access is often directly granted from the user when he/she is using a particular service for browsing particular sites. In many cases, a lot of this private information is captured by online services even when the direct consent or knowledge of the user is not there. So their party services follow users in order to track the users across the different websites which they access. When a user surfs the web, they leave traces of their identity in the form of the patterns of their activities and many more such unstructured data which creates the users' online footprints (Acar et al. 2013, 2014). The fact that users carry around a host of personal computers and other communication devices makes them locatable, identifiable, and trackable across different locations, networks, and services. Therefore, this information, arising from users' activities, along with other technologies, can effectively enhance the surveillance capabilities and lead to an effective monitoring system in the interest of national security. Figure 11.1 shows how service providers track the chain of websites browsed by a user, to provide advertisements the user may be interested in.

Recommender systems (Adomavicius and Tuzhilin 2005; Ahn 2006; Bailey 2008) are software tools that use agents to help the user to find the most suitable pages of their interest. The algorithms used by the recommender systems are mainly of three types: collaborative filtering, content-based, and hybrid methods. In a *content-based system*, recommendations are made by collecting information about the profile features of a user. The idea is that if a user has shown interest in the past for a particular thing, then it is very likely that the user will also be interested in that object again in the future. Usually, objects of similar types are put in a group based on the similarity of their features (Balcan et al. 2006; Boutilier et al. 2003; Bridge and Ricci 2007). The profiles of users are created either by the use of historical interactions or by

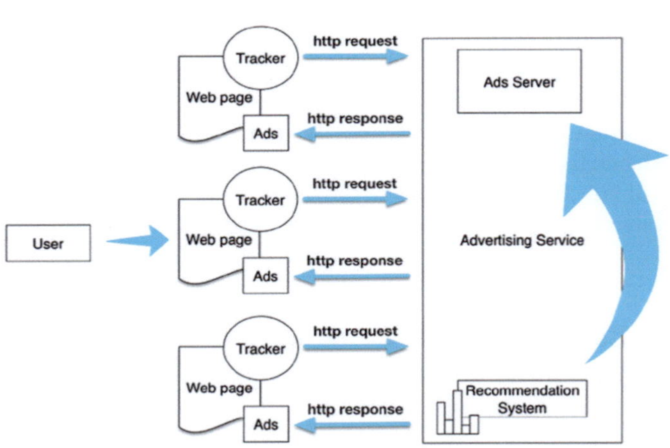

Fig. 11.1 Advertisers respond with corresponding advertisements based on user web search activity (Puglisi et al. 2017)

explicitly asking users about their interests. Figure 11.1 explains how recommender systems work.

In a collaborative filtering system, user interactions are utilized to filter out the objects which are of interest. The set of interactions is visualized as a matrix, where the interactions between the users *i* and items *j* are represented by the entries (*i*, *j*). It can be thought of as a generalization of classification and regression. But while in the former case it is predicted whether a variable directly depends on the other variables, in collaborative filtering no such distinction is made between the feature variables and the class variables. The problem is visualized as a matrix, but instead of predicting the values of a unique column, the values of any given entry are predicted. At present, this is one of the approaches being used most frequently and normally provides results that are better than content-based recommendations. The recommendation systems of YouTube, Netflix, and Spotify use this type of system (Box et al. 2005; Breiman and Breiman 1996; Puglisi et al. 2017).

Hybrid systems are a combination of both types of information and are aimed to overcome the issues that come up while working with just one kind.

11.4 Sparse Matrices

In general, most of the large matrices are found to be sparse in nature. A recommender system matrix also is usually a sparse matrix. A matrix as we know is a two-dimensional data object made of m rows and n columns, so the total number of values is *m x n*. If most of the elements of the matrix have a "0" as a value, then the matrix is called a *sparse matrix*. The advantages of a sparse matrix over a normal dense matrix are as follows:

Storage: The number of nonzero elements is lesser than the number of elements, and therefore, the amount of memory needed to store only the nonzero elements will be lesser.

Computing time: The computing time may be reduced by the logical design of a data structure that traverses only the nonzero elements.

Suppose we take the following sparse matrix in Fig. 11.2.

If we represent it as a 2D array, then it will waste a lot of memory, as the zeros in the matrix are usually not needed in most instances. Therefore, instead of storing the zeroes also, we store only the nonzero elements, i.e., as a triple (row, column, value).

Fig. 11.2 Sparse matrix

0 0 7 0 5

0 0 3 8 0

0 0 0 0 0

0 4 2 0 0

Fig. 11.3 Array representation of a sparse matrix

Row	0	0	1	1	3	3
Column	2	4	2	3	1	2
Value	7	5	3	8	4	2

Although sparse matrices can be represented in various ways, the two most frequent ways to store them are as:

a. Array representations.
b. Linked list representations.

In the array representation, the sparse matrix is represented as a 2D array whereto the following three rows are used:

Row: Index of a row, this is where the nonzero elements are located.

Column: Index of a column, this is where the nonzero elements are located.

Value: This is where the value of the nonzero elements is located at index (row, column).

So, the above matrix is represented as follows in Fig. 11.3.

In the linked list representation, each node contains four fields. These fields are defined as:

Row: This is the index of the row, which contains the location of the nonzero element.

Column: This is the index of a column, which contains the location of the nonzero element.

Value: This is the value of the nonzero element which is located at the index (row, column).

Next node: This contains the address of the next node.

11.5 Our Proposed Framework

The amount of information that can be extracted for surveillance is difficult to determine because its accuracy depends on four factors: the web structure, how the web resources are mapped to the topology of the global Internet, a typical user's web-browsing behavior, and the technical capabilities and the policy restrictions of the adversary/authority.

The data for the tracking reports comes from the three following sources:

- The HTTP request of the user.
- The browser/system information.
- First-/third-party cookies.

So the tracker is capable of inspecting the packet contents and can either do the tracking of an individual target or surveil a large group of users. Although a major challenge here is the dearth of persistent device identifiers, this can be overcome to a large extent by observing third-party cookies. Since there are multiple unrelated third-party cookies on most web pages, they can be tied together to most of a user's web traffic, even without the IP address. So the network traffic can be separated into clusters, where each cluster corresponds to only one user.

The following are the steps for the targeted surveillance process:

- Either the target identity is scanned in plaintext HTTP traffic or some auxiliary method is used to get the targets cookie ID on some first-party page.
- Then, the target known first-party cookie can be transitively connected to other third-party and first-party cookies of the target. In the case of en masse surveillance, all the HTTP traffics can be clustered first, and then individual identities attached to these clusters.
- Once the identities have been attached, various types of information and activities can be extracted or predicted. Firstly, the browsing history itself may provide primary information about the interests of the user, e.g., terror attacks. Secondly, it can also provide sensitive information in unencrypted web content like purchase history, address, etc.

Our proposed framework tracks the web browsing of users to build a sparse matrix similar to a recommendation system, based on the tracking patterns. The values of the matrix and the sites are monitored to see if a user is suspected to traverse the Internet with malicious intentions. Then an alert system is in place to raise flags to the concerned authorities.

Figure 11.4 shows the process of tracking a user's browsing information. The user sends HTTP requests to various websites. All these websites will now tell the user to browse to send a request to the same tracker for an ID. The user's browser will now send requests to the tracker, and the tracker replies with an instruction to set a cookie with an ID. For single website trackers, it is a first-party cookie, while cross-website trackers will store their ID in a third-party cookie. So in this case the tracker will set a single third-party cookie with an ID that can be accessed over all of the websites. This information from the trackers' set for multiple uses can be used to build and populate a recommender system sparse matrix. When the number of hits to a particular page that has been flagged as potentially dangerous, which goes beyond a particular threshold value, then it sends an alert to the surveillance system administrator.

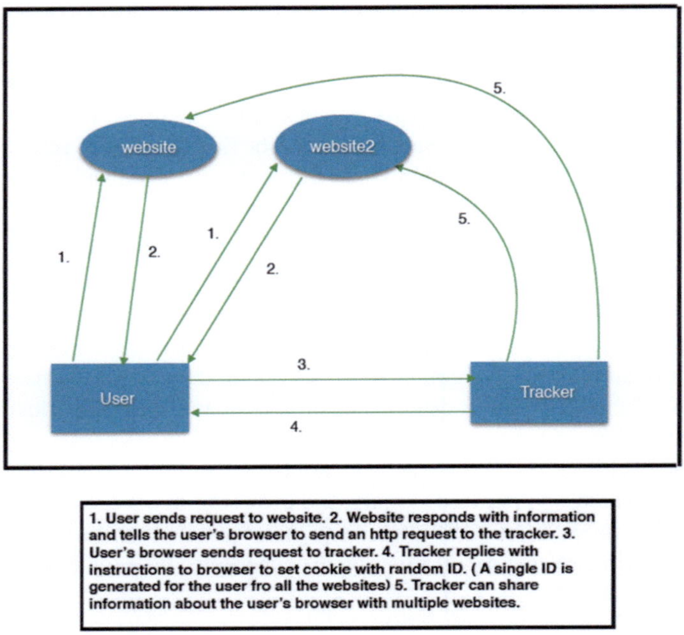

Fig. 11.4 Tracking browsing patterns of users with a third-party tracker

11.6 The Proposed Algorithm for the Threat Analysis and Alert

A list of users and the list of websites browsed by them are tracked using a third-party tracker using the following algorithm. The browsing data and other related information are shared by the tracker to build the recommender alert system matrix. Whenever access to harmful sites is detected in the matrix, it will send alerts to the concerned authorities.

```
FindThreat( )
  Input : userIDList[] ;
          websiteList[];
          cookieID[];
          trackers_3P[];
          threshold_value;
  Output: threatAlert( );
          user_ID;
  for each user in userIDList[] do
     for each website in websiteList[] do
          send httpRequest() to each website in websiteList[];
          website response with webpage data and tells browser to send cookie_ID request
to browser;
          browser requests for cookie_ID[]to tracker_3P[];
          tracker  response with cookieID[] to user[];
          generateCookieID( );
          share user browser data with website;
   for i=1 to n do
      for j=1 to m do
         buildRecommenderSystem( ) R[i][j];
      end for
   end for
      end for
  end for
   for i=1 to n do
      for j=1 to m do
        if R[i][j]>= threshold_value, then
              send threatAlert(), user_ID;
      end for
    end for
```

11.7 Real Example

A real example of Surveillance Recommender System is available in GitHub repository. Readers can access the source code of the project through the following link:

https://github.com/pushpendukar/Surveillance_RS.git.

11.8 Summary

In this chapter, we have proposed a framework for the surveillance and tracking of suspicious browsing activities by a user. It maintains a sparse matrix of the chain of websites browsed by a user as well as a list of probable sites they may access. Based on this sparse matrix, if the number of hits to harmful crosses a threshold value, then an alert is sent to the authorities. It can also be an effective tool for the threat analysis for potentially malicious users on the website. In the future, we plan to extend this work to deal with cases where the users are browsing in incognito mode.

Think Tank

1. *What is cyber-surveillance and why is it important?*
2. *What is a cookie?*
3. *What is a sparse matrix?*
4. *How can we user recommendation systems for surveillance?*

References

Acar G et al (2013) FPDetective, dusting the web for fingerprinters. In: ACM SIGSAC 2013 conference on computer & communications security, pp 1129–1140

Acar G et al (2014) The web never forgets: persistent tracking mechanisms in the wild. In: 21st ACM conference on computer and communications security (CCS 2014)

Adomavicius G, Tuzhilin A (2005) Toward the next generation of recommender systems: a survey of the state-of-the-art and possible extensions. IEEE Trans Knowl Data Eng 17(6):734–749

Ahn LV (2006) Games with a purpose. Computer 39(6):92–94. https://doi.org/10.1109/MC.2006

Bailey RA (2008) Design of comparative experiments. Cambridge University Press

Balcan MF, Beygelzimer A, Langford J (2006) Agnostic active learning. In: ICML'06, 23rd international conference on machine learning. ACM, New York, NY, USA, pp 65–72

Boutilier C, Zemel R, Marlin B (2003) Active collaborative filtering. In: 19th annual conference on uncertainty in artificial intelligence, pp 98–106

Box G, Hunter SJ, Hunter WG (2005) Statistics for experimenters: design, innovation, and discovery. Wiley-Interscience

Breiman L, Breiman L (1996) Bagging predictors. Machine learning, pp 123–140
Bridge D, Ricci F (2007) Supporting product selection with query editing recommendations. In: RecSys'07: proceedings of the 2007 ACM conference on recommender systems. ACM, New York, NY, USA, pp 65–72. https://doi.org/10.1145/1297231.1297243
http://donottrack.us/
http://trackingobserver.cs.washington.edu (2014) TrackingObserver: a browser-based web tracking detection platform
https://www.ghostery.com/
https://sharemenot.cs.washington.edu (2014) ShareMeNot: protecting against tracking from third party social media buttons while still allowing you to use them
Puglisi S, Monedero D, Forne J (2017) On web user tracking of browsing patterns for personalised advertising. Int J Parallel Emergent Distrib Syst. https://doi.org/10.1080/17445760.2017.1282480

Chapter 12
Some Novel Applications of Recommender System and Road Ahead

Abstract This chapter gives a brief introduction to the recommender system and provides details of six different applications of the recommender system (health care, security, tourism, e-commerce, e-learning, and social network). It mainly discusses reasons to use recommender systems, real-life examples, as well as techniques such as collaborative filtering, content-based filtering, and hybrid filtering behind it. Based on the existing problems within each recommender system, some possible solutions are given to improve the current recommender systems, respectively.

Keywords Recommender systems · Security · Tourism · E-commerce · E-learning · Social network

12.1 Introduction

Nowadays, the rapid development of the Internet enables a large influx of information provided for users through networks and different platforms. However, this leads to information-overloaded problems, which means that users cannot find exactly what they really want in a short time due to a large number of dazzling choices, making them hard to make decisions (Kunaver and Požrl 2017). Therefore, in order to deal with this problem, it is necessary to come up with solutions to help users quickly find out what they want and provide them with the most appropriate products or services (Lu et al. 2020). That is how recommender systems came into being to solve this problem (Lu et al. 2015). Recommender system (RS) is a filtering system that uses complex algorithms to select the most relevant items, content, or services that users most frequently search for, tailored to their preferences (Das et al. 2016; Isinkaye et al. 2015). It has now been widely used in all parts of our life, such as e-commerce, traveling, and health care. For example, Amazon uses recommender systems to help users find books or other products they like. Moreover, introducing recommender systems not only makes life much easier for users but also benefits companies and providers. This documentation is divided into three parts: algorithms (techniques,

P. Kar et al., *Recommender Systems: Algorithms and their Applications*, Transactions on Computer Systems and Networks, https://doi.org/10.1007/978-981-97-0538-2_12

models) behind recommender systems, applications of recommender systems, and new areas where recommender systems can be implemented.

12.2 Applications of Recommender System

12.2.1 Security

12.2.1.1 Why Do We Need to Use Recommender Systems in Security?

To realize the personalized recommendation system, and make its recommendations more accurate, the user's sensitive data (such as sex, occupation, location, and age) are needed partly acquired by a machine. Therefore, there is the possibility of sensitive data leakage in the user data collection, storage, and transmission in the process. In this case, it is necessary to put more emphasis on user privacy protection and security recommendation system (Weiming et al. 2019).

The first obvious reason why a recommendation system is applied in the field of security is that in the process of personalized recommendation on web pages, which is becoming increasingly popular today, the system will inevitably extract users' personalized information (such as social tags and personal preferences), which will lead to threats to the privacy of user profile. However, adopting adequate privacy protection policies may reduce the quality of user-recommended content (Katarya and Verma 2017). Therefore, the unique point about the application of recommendation systems in privacy protection is to discuss the premise of ensuring the quality of recommendation, while taking user privacy into account.

The second reason is that the application of recommendation systems in security is beneficial for early detection and warning of victims. For all kinds of network attacks that violate privacy, a recommendation system can analyze and study all kinds of malicious behavior patterns of privacy disclosure, so as to formulate corresponding strategies.

12.2.1.2 Techniques Used in Recommender Systems for Security

- Collaborative filtering (CF)

In daily life, people tend to consult their friends or trust for an unfamiliar problem or something and make their own choices based on these judgments and opinions. A typical collaborative filtering algorithm is based on the user's collaborative filtering algorithm. Its basic principle is to get user neighbors using historical rating data and recommend to the target user according to rates similar to the nearest neighbor of the score data (Yubo et al. 2010).

Therefore, the unrated projects' score target user-made can predict through the nearest neighbor approximation of the weighted average, to produce recommendations. This process comprises three steps to complete: data presentation, finding the nearest neighbor, and producing recommendations (Yubo et al. 2010).

This recommendation method involves the use of k-nearest neighbors. KNN algorithm is used to reduce the exposure of user privacy and protect user files from privacy threats based on not reducing the quality of the content recommended to users (Katarya and Verma 2017).

Web service recommendation methods are also based on collaborative filtering recommend services and Mashup call history, user similarity, or service similarity. This method is generally used for Quality of Service (QoS) prediction in the early days, selecting high-quality services in web service recommendations.

- Statistics-based recommendation

A statistical method is a mature privacy protection method used in the data calculation stage (such as similarity calculation). It can process sensitive information in data files by removing features, obfuscation, adding noise, and other methods. This is often referred to as the anonymization algorithm. At present, K-anonymity, I-Diversity, and T-Total are the most common authorization and acceptance methods (Weiming et al. 2019). Although these three statistical methods are very efficient and relatively simple to calculate, sometimes they do not play a substantial role in user privacy theft with common characteristics (Sweeney 2020).

For example, if the dimensionality of information obtained from users is sufficient, even though the use of these technologies anonymizes a single column of information, the attacker can still compare and categorize, retrieve all the user's information, or relocate the user personally (Wong et al. 2006).

- Cryptography-based recommendation

This mainly includes homomorphic encryption technology; homomorphic encryption technology means that we can directly perform some operations on ciphertext so that it can still achieve the same effect on the plaintext after the same operation and does not affect the confidentiality of the plaintext. The core technology is multi-secret computing, which is applied to a variety of recommendation systems, such as neighborhood-based collaborative filtering, machine learning-based collaborative filtering, and content-based recommendation. Currently, cryptography-based privacy protection methods have advantages in security but consume more computing resources, which cannot meet the immediacy of large recommendation systems (Weiming et al. 2019).

- Hybrid filtering

Another approach to filtering techniques is a combination of the two basic filtering techniques, content-based and collaborative (Allen 2019). This technique is referred to as the hybrid filtering technique which is another popular approach to filtering. The hybrid technique combines two or more filtering methods. It thus gains an increase in accuracy and performance.

- User identification

Since the existence of proxy servers and firewalls, user identification cannot be distinguished by IP address. Then, we should adopt the heuristic rules. Different IP addresses represent different users. When the user's IP addresses are the same, different operating systems or browsers can represent different users. When the user's IP addresses, operating systems, and Internet explorers are the same, website topologies represent different users. When a web cannot be reached from the history of previously visited pages, then it is considered a new user (Zhang and Wang 2013).

- Session identification

Session identification is to break records of web pages that the user has visited into separate sessions. It considers all the web pages that a user visited at one time as one user session. Because it is difficult to determine whether the user has left the website, the easiest way to determine whether the user has left the site is to use the maximum timeout. If the period between two pages' request is more than a certain time limit, then we consider that the session has been finished, then start a new session. A large number of statistical studies have shown that 30 min is relatively standard for judging time out or not (Zhang and Wang 2013).

12.2.1.3 Scope of Improvements in Recommender Systems for Security

- The matrix factorization algorithm cannot effectively capture the complex interaction information between the two in the sparse Mashup–web service call matrix, which may potentially result in lower recommendation performance.

12.2.2 Tourism

12.2.2.1 Why Do We Need to Use Recommender Systems in Tourism?

Nowadays, recommender systems have been successfully applied in many fields. However, there are still a lot of things to improve the recommender systems in tourism (including e-travel) (Bayati et al. 2022). Moreover, as there is such information online, it would be difficult for tourists to find the most accurate information about the places they want to go to in a short time (Fararni et al. 2021). Therefore, by using a recommender system in tourism, the decision-making process can be facilitated (Sondess et al. 2019).

12.2.2.2 Real-Life Examples of Recommender Systems in Tourism

These three websites (https://www.ctrip.com, https://www.mafengwo.cn, and http://www.tripadvisor.com) as well as their corresponding mobile applications use a recommender system to help provide a better user experience. They can gather the information through the users' browsing history and by using content-based filtering as well as other strategies, find similarities between the target user and other users in the database.

In the tourism domain, many recommender systems have been developed. Most of them are content-based and knowledge-based and confined to producing recommendations at the destination level. For instance, Triplehop's TripMatcher and VacationCoach's Me-Print use the content-based method to make recommendations on destinations. The knowledge-based systems deduce the preferences and requirements of the tourist based on the knowledge obtained from the conversational dialog.

Another example is Tripadvisor. The advantage of this type of system is that it only uses the demographic data of the user such as gender, age, and education and may not need the history of users' ratings, the textual description, or the knowledge of items. Therefore, new users can get recommendations before they rate any items.

12.2.2.3 Techniques Used in Recommender Systems in Tourism

- Collaborative filtering (CF)

This is about making a recommendation to the target user based on items with high ratings made by users who have similar preferences to the target user (Fararni et al. 2021). Nowadays, more factors can be taken into consideration, such as weather condition and location (latitude, longitude, and GPS coordinates) as shown in Fig. 12.1.

Fig. 12.1 Collaborative filtering in tourism recommender system (Sondess et al. 2019)

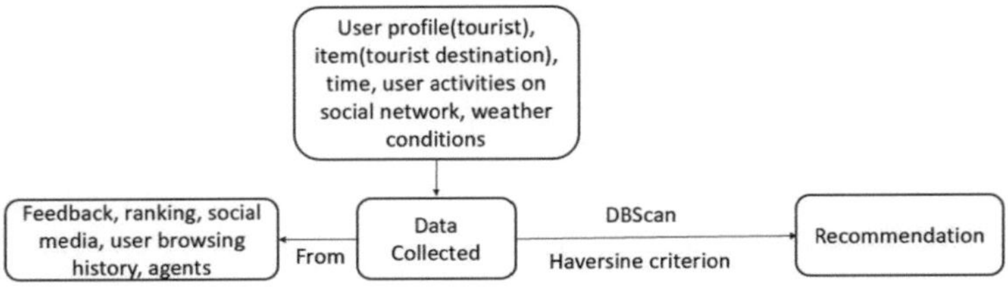

Fig. 12.2 User-generated content in tourism recommender system (Bayati et al. 2022)

In addition, user-generated content (UGC) is used to help make a better recommendation to the target user. The information is mainly from the social media that the target user uses (Sondess et al. 2019).

In order to improve collaborative filtering to make a better and more accurate recommendation, DBScan clustering and Haversine similarity criterion can be combined in collaborative filtering. DBScan algorithm can be introduced to cluster user datasets and items. Then, by using the Haversine similarity criterion, similar items can be recommended to the new user more accurately by deciding the class that the new user belongs. This is based on k similar users to the new user and items with higher rates (Bayati et al. 2022). Therefore, the recommendation is more accurate.

Figure 12.2 shows how the collaborative system works with DBScan and Haversine similarity criterion.

- Content-based filtering

This is based on the user preferences and similarity analysis between items and then recommending to the target users (Fararni et al. 2021). Multi-user profiles and TR-service profiles can be used in this algorithm (shown below).

In order to provide the best recommendation to the target user, the target user's profile must be compared with the profile of other users in the TR service. Therefore, it is also important to calculate the similarity between the distribution of the target user, θ_c, and the distribution of the user profile layer in the TR service, denoted by θ_s, which is positive feedback from other users. The relevance score is calculated by Eq. 12.1.

$$\hat{r}_{u,s,c} = \frac{1}{D_{KL}(\theta_c||\theta_s)}, \tag{12.1}$$

where $D_{KL}(\theta_c||\theta_s)$ refers to the divergence between these two probability distributions, which can be interpreted as Eq. 12.2:

$$D_{KL}(\theta_c||\theta_s) == P(w|R_s) \log \frac{P(w|R_s)}{P(w|R_c)}. \tag{12.2}$$

It should be noted that $D_{KL}(\theta_c||\theta_s) \neq 0$ for each s.

The higher the score, the more similar is between the target user and users in the TR service. Therefore, recommendations can be made more accurately based on the relative score (Sondess et al. 2019). However, it is not logically sound since the target user may not want to go to the same place twice (Fararni et al. 2021).

- Hybrid filtering

The idea of this approach is to maximize the potential of collaborative filtering and content-based filtering (Fararni et al. 2021), making the recommender system more robust. Big data systems, deep learning techniques, and social networks can be used to help make the filtering algorithm and recommender system more robust.

- Knowledge-based

In comparison with collaborative filtering and the content-based recommender, the knowledge-based recommender collects information specifically for the domain of travel destinations and is thereby not directly applicable to other domains. The following information sources were integrated into the knowledge-based recommender: (1) geographic information: the exact location (longitude and latitude), continent, and country of each destination. (2) Travel costs: the costs of traveling from your current location to the destination in question. (3) Attraction types: what specific attraction types can be found at that destination? (4) Tourist profile (stereotypes such as backpackers, family Travelers): how each location matches typical tourist profiles as defined in Gogobot (Zheng et al. 2018).

12.2.2.4 Scope of Improvements in Tourism Recommender Systems

- Collaborative filtering is only able to make recommendations based on the existing ratings by previous users. However, if there is a new item since there is no rating history about this item or there is a new user, an accurate recommendation cannot be provided (Bayati et al. 2022).
- Though there is a hybrid technique in the recommendation approach, more than 90% of the applications in tourism just use a single technique instead of using the hybrid one to make recommendations (Fararni et al. 2021).
- Sometimes the recommendation is not accurate or logically sound since the target user may not want to go to the same place twice (Fararni et al. 2021).
- Recommendation to the new user is not accurate since there is no previous information about the new user.
- The user's preference is not stable. It may change as time passed (Zheng et al. 2018).

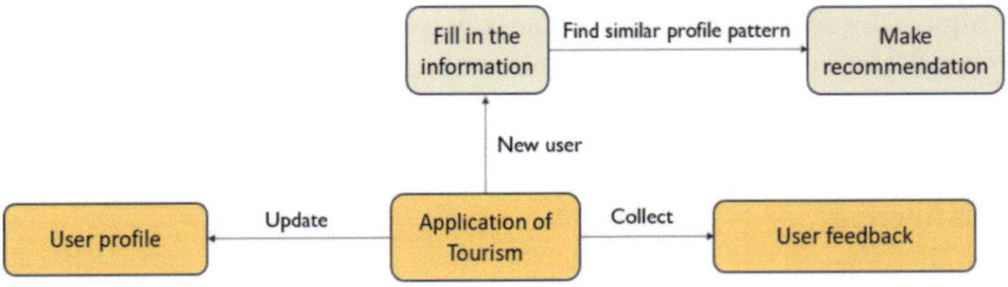

Fig. 12.3 Improvement in tourism recommender system

12.2.2.5 Possible Ideas for Improvement of the Existing Tourism Recommender Systems

- Collect user feedback frequently to see if there are things that need to be improved. For example, we can collect feedback from the user once a month to see if there are things that need to be improved.
- Ask the new user to fill in his/her information before browsing the web pages about tourism. Then by building the user profile, find a similar user profile pattern in the database and make recommendations based on this.
- As Fig. 12.3 shows, updating the user profile frequently to keep track of the user's latest preference and make a more accurate recommendation.

12.2.3 E-commerce

12.2.3.1 Why Do We Need to Use Recommender Systems in E-commerce?

As the Internet and smart mobile devices evolve rapidly, online shopping has become an indispensable part of our lives. It seems to benefit us a lot and make our life much easier since we can buy things that we want without stepping out of our homes. However, the great number of items makes it hard for the customer to make choices, which may discourage customers from buying reliable products, thus hurting the economy (Duo and Su 2015). In addition, companies rely on the users' shopping data to make analyses (Hidayatullah and Anugerah 2018). Therefore, recommender systems in e-commerce are introduced and needed to increase customers' online shopping experience by using complex algorithms to make accurate recommendations to customers and satisfy their requirements (Fu and Leng 2018).

12.2.3.2 Real-Life Examples of Recommender Systems in E-commerce

Taobao (https://www.taobao.com) and Amazon (https://www.amazon.cn) as well as their corresponding mobile applications use recommender systems to help provide a more accurate recommendation to customers by using different algorithms.

Amazon's recommendation algorithm is probably the most complex and effective in the e-commerce market. Its recommender system is capable of intelligently analyzing and predicting customers' shopping preferences in order to offer them a list of recommended products. Amazon's algorithm selects recommended products for each user based on their previous purchases, interactions, and ratings of other items on display and combines them with similar items viewed by users with similar preferences and interests. Many of these systems can be classified as content-based filtering or collaborative filtering. The strategies listed above are currently being developed and refined by the Amazon team in order to deliver the best product recommendations to customers.

12.2.3.3 Techniques Used in E-commerce Recommender Systems

- Collaborative filtering (CF)

This technique is one of the most popular techniques that are used in e-commerce. It can be divided into a user-based clustering model and an item-based clustering model (Zhao and Ji 2013). It focuses on the other users who share similar shopping behaviors (i.e., buying the same product) with the target user (Ouaftouh et al. 2019). Similarity calculation is used in this technique to make the recommendation more persuasive. It is described as Eq. 12.3.

$$\text{sim}(\vec{x}, \vec{y}) = \cos(\vec{x}, \vec{y}) = \frac{\vec{x} \cdot \vec{y}}{||\vec{x}|| * ||\vec{y}||}. \tag{12.3}$$

Cosine similarity is used to find the similarity between two users x and y (Zhao 2019). Therefore, it can effectively filter the user that does not share any shopping similarities with the target user. Then, the recommender system will make a recommendation about items that similar users have bought to the target user that he/she did not buy. This can indefinitely enhance the accuracy of the recommender system.

- Content-based filtering

Products' attributes are analyzed before performing content-based filtering.

For example, if a customer bought a scientific book with a high rating score, then the system will make a record of this, as well as the rating made by the customer, then next time the recommender system will make a recommendation to the customer with books of the same genre as shown in Fig. 12.4.

Association rule mining is one of the commonly used algorithms in content-based filtering, which can find interesting pattern between variables in large hidden

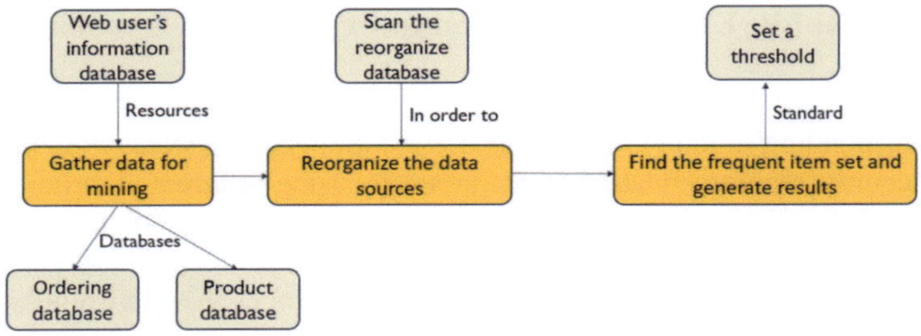

Fig. 12.4 Content-based filtering in e-commerce recommender system

datasets. Association rules or frequent items can be used to describe the relationship (Zhao 2019). Two steps are needed to achieve association rules. Firstly, a minimum threshold probability for the item set. Let us say minPro $= m$, any frequent item set with a probability greater than m can be chosen. Secondly, assume the confidence of the item is c, and X and Y are two frequent item sets ($X \cap Y = \varnothing$). If $P(Y|X) \geq c$, then $X \rightarrow Y$ can be classified as the association rule as.

- Hybrid filtering

The idea of this approach is to maximize the potential of collaborative filtering and content-based filtering (Fararni et al. 2021), making the recommender system more robust. Big data systems, deep learning techniques, and social networks can be used to help make the filtering algorithm and recommender system more robust.

- User clustering models

Clustering, along with other unsupervised learning, helps to make recommendations more accurate and reliable. A classic use of the clustering method is the K-Means method, which always tries to cluster the nodes with the same center node by calculating the Euclidean distance until it is stable. Combining the K-Means method with e-commerce, this technique is used to cluster customers with similar shopping behavior and user profile (Ouaftouh et al. 2019).

This technique seems to be more efficient than the collaborative filtering technique since the comparison is made only within the clustering group, rather than comparing with all users in collaborative filtering (Zhao 2019).

The vectors of different users can be formed after clustering (Zhao 2019). The expression is shown in Eq. 12.4:

$$\text{UserProfile} = \{(a_i, q_i), i \in \{1, \ldots, n\}\}, \tag{12.4}$$

where user refers to the user profile attribute, and q refers to the corresponding value of the attribute.

Another similarity can be used to calculate the similarity between users U_1 and U_2 in the same clustering group (shown in Eq. 12.5):

$$\text{similarityInSameGroup}(U_1, U_2) = \sum w_i \times \text{sim}_i(x_i, y_i), \tag{12.5}$$

where w_i refers to the weights of different attributes in the user profile and $\text{sim}_i(x_i, y_i)$ is calculated based on a similarity metric (Zhao 2019).

12.2.3.4 Scope of Improvements in E-commerce Recommender Systems

- There are some existing gaps in collaborative filtering. For example, some users would not like to rate or make comments on the items they bought. This may affect the user profile and similarity value, thus lowering the accuracy of the recommendation. Another example is that the cold start is still to be solved, which means that when a new user is shopping on the e-commerce website for the first time, there will be no recommendation to the user because there is no shopping history for this user (Zhao 2019).
- Although collaborative filtering is successfully used most of the time, there still exist some potential situations that need to be considered. For example, customers may change their preferences over time (Zhao 2019).
- Even though some recommender systems can provide users with relatively accurate recommendations, the quality of the product cannot be guaranteed since some users are employed by the shop to make a high rating for its products to make more people browse his shop and buy products in his shop.

12.2.3.5 Possible Ideas for Improvement of the Existing E-commerce Recommender Systems

- If a new user logs in to the shopping app for the first time, we may list a variety of style of clothes and daily necessities and then let the user chooses his/her preferences based on these. Also, if there is something not listed in the interface, we may give the user a chance to type in what he/she may buy or may be interested in. Thus, after the user enters the app, items according to his/her preferences are recommended. This might solve the cold-start problem.

For example, if a user uses Taobao for the first time, he/she should be given a list of items that he/she might be interested in, such as clothes and shoes. If the items are not given in the list, the user should also be able to write down the items that he/she would like to buy, such as watches. This is presented in Fig. 12.5.

Fig. 12.5 Improvement of recommender system in e-commerce

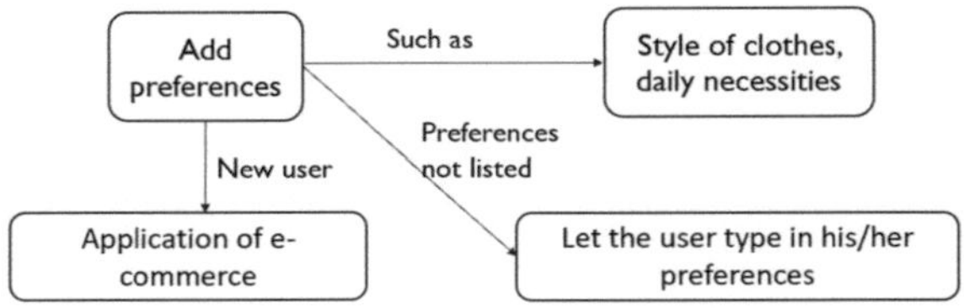

12.2.4 E-learning

12.2.4.1 Why Do We Need to Use Recommender Systems in E-learning?

With the advent of more advanced technology, individuals can learn online instead of offline, making individuals' timetables more flexible. E-learning systems are around to provide students and learners with virtual educational environments in which they do not need others' assistance in the process of learning. Through e-learning, they can have access to a wide variety of learning resources (Ansari et al. 2016). However, due to information overload, many learners are experiencing challenges in retrieving relevant and useful learning resources that meet their needs. It seems that the core component of a working and efficient e-learning system is its recommender system (Ansari et al. 2016). The recommender system in an e-learning context tries to intelligently recommend learning resources to a learner based on the task already done by the learner (Nath 2018).

12.2.4.2 Real-Life Examples of Recommender Systems in E-learning

MOOC (http://www.icourse163.org) is an online learning platform based on clustering and course-based collaborative filtering. It makes a recommendation based on other users that have similar interests and hobbies to the target user or other courses that are like the target user's historical data. It also considers the situation of adding a new user and new courses which is an extremely difficult and practical problem when building the recommender system in the real world. For new users, because they do not have any record of accessing courses, the collaborative recommendation algorithm can no longer work. For these users, the framework makes the recommendation mainly from two aspects: recommendation based on user profile and recommendation based on hotspot courses. Coursera (http://www.coursera.org) follows this technique.

12.2.4.3 Techniques Used in E-learning Recommender Systems

- Collaborative filtering (CF)

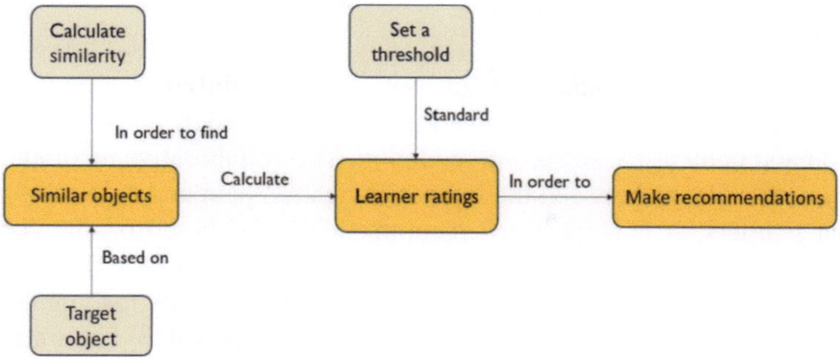

Fig. 12.6 Collaborative filtering in recommender system in e-learning (Isinkaye et al. 2015)

This technique aggregates the rating of objects to recognize commonalities between learners and generate new recommendations based on interlearner comparisons. A learner profile consists of a vector of learning objects and their ratings. Ratings indicate the degree of preference. It may be binary (likes/dislikes) or real-valued. It can be further divided into memory-based and model-based.

The memory-based technique can be classified into user-based and item-based. The user-based model is as follows: Each learner resides in a group of similarly behaving learners and finds a set of learners with similar preferences. Finally, it generates a list of recommendations for the target learner. The item-based model identifies the set of the learning object that is similar to or related to the target learner's liked objects. After that, it computes the similarity of learning objects and finds the most similar objects to the target objects within the set of learning objects that the learner has rated. Model-based techniques provide recommendations by estimating statistical models for learner ratings. A probabilistic method can be used to compute the probability that the learner gives a particular rating to a new learning object based on previously rated objects (Nath 2018). This is presented in Fig. 12.6.

- Content-based filtering (CBF)

This technique is based on a comparison of the content of the learning object and a learner profile. The two classes of content-based recommendation are case-based reasoning techniques and attribute-based techniques. A case-based reasoning technique recommends learning objects that are in highest correlation to objects the learner liked in the past. This technique does not desire a content analysis. The quality of the recommendation rises when the learners have rated more learning objects. The new learner problem also stated case-based reasoning techniques. The limitation of this technique is overspecialization because it recommends only the learning objects that are in higher correlation with the learner profile or interest. In attribute-based techniques, learning objects are recommended based on the mapping of their attributes to the learner profile. Attributes could be weighted for their relevance to the learner. This technique is sensitive to changes in the learner profile (Nath 2018).

- Hybrid filtering

Hybrid filtering is a collaboration of two or more different recommendation approaches. The widely known hybrid approach is provided by collaborative recommendation and content-based recommendation. The collaborative recommendation is based on a similarity between the learner navigation path and the access patterns of similar learners. A content-based recommendation is based on the correlation between the content of the learning objects and the learner's taste. Hybrid recommendation tries to overcome the limitations in each approach, by making the collaborative recommendation deal with any type of content and explore a new area to find something that is interesting to the learner (Nath 2018).

- Knowledge-based recommendation (KB)

This recommender system attempts to propose objects based on a learner's needs and preferences. It contains knowledge about how a specific learning object meets a specific learner's needs. This technique collects knowledge about the learners and learning objects to apply them to the recommendation activity. It is independent of learner ratings. For example, cars may have several models, colors, engine options, and interior options, and user interests may be regulated by a very specific combination of these options. It does not collect data about a specific learner because its intuition is independent of individual preferences. Knowledge-based techniques are suitable for hybridization with other recommendation techniques in the case of e-learning recommenders (Nath 2018).

- Demographic recommendation

This technique classifies the learners based on their personal attributes and the recommendations are based on the demographic classes. This approach assumes that all learners belonging to a certain demographic class have alike interests or preferences. It uses demographic data about the learner and their point of view for the recommended learning objects. It forms people-to-people correlations like collaborative ones. But they use different data, such as the same age group. The benefit of this approach is that it is independent of learner rating history (Nath 2018).

- Context-aware systems

  (a) Traditional recommender systems compromise with two types of entities, users, and items. The recommender system includes additional information about the learner's context such as available time, location, and people nearby. These data can be used to change recommendations based on the individual learner's characteristics. Context is information that can be used to classify the situation of an entity. An entity is an object, person, or place that can be considered relevant to the interaction between an application and a user. The context data consists of different attributes, like physical location, date, season, emotional state, physiological state, personal history, etc. For example, a website may recommend songs to a user by asking about the current mood of the user. This system automatically uses context data to run the system that is suitable for a specific time, place, or event. It is necessary to combine the context data into the recommender systems to recommend learning objects to the learners under some circumstances. It covers the understanding of the learner's objective with objects that learners might find interesting by knowing the wide area of contextual attributes (Nath 2018).

12.2.4.4 Scope of Improvements in E-learning Recommender Systems

- When there is a new learner to the system who has no prior rating found in the rating table, it is difficult to give a prediction of a learning object for the new learner because it requires the learner's historic rating to calculate the similarity for determining the neighbors.
- A cold start problem for a new learning object occurs when there are not enough previous ratings related to that learning object exists (Nath 2018).
- Data sparsity occurs when the number of learners who have rated a learning object is too small compared to the number of available learning objects. If there is no such overlap in ratings with the target learner occurs, it is difficult to generate appropriate recommendations (Nath 2018).
- Specialization is the major problem faced by the content-based recommender system. The learners are recommended with learning objects that are already familiar with. It prevents learners from finding new learning objects and other alternatives. Additional techniques must be added to the system to make suggestions outside the scope of learner interest. By integrating additional methods, the learner will be provided with a set of different and wide ranges of options (Nath 2018).
- In the context of a demographic recommender, privacy is considered to be a major issue. To provide a more accurate recommendation to the learner, the most sensitive data of a learner must be acquired. It includes demographic information and information about the location of a specific learner, which may rupture the privacy of the learner (Nath 2018).

12.2.4.5 Possible Ideas for Improvement of the Existing E-commerce Recommender Systems

Collect user feedback frequently to see if there are changes in the user's needs and then changes need to be made accordingly. For example, if a user does not like fashionable clothes anymore, then the app should reduce the recommendation of fashionable clothes.

12.2.5 *Social Network*

12.2.5.1 Why Do We Need to Use Recommender Systems for Social Networks?

Nowadays, many people use social media to communicate with others, share their interests, and obtain information. Recommender systems are surely the applications that can take the most immediate and evident advantages by leveraging in different ways user experiences and interactions within a social community to suggest multimedia objects of interest (Amato et al. 2017).

Nowadays, recommender applications and services have been introduced to support effectively and efficiently the intelligent browsing of items' collections, assisting users to find "what they need" within this ocean of information and thus realizing the well-known transition in the web from the "search" to the "discovery" paradigm (Amato et al. 2017).

12.2.5.2 Real-Life Examples of Recommender Systems in Social Networks

Just as a real example, each minute thousands of tweets are sent on Twitter, several hundreds of hours of videos are uploaded to YouTube, and a huge quantity of photos are shared on Instagram or uploaded to Flickr.

In the case of image-sharing social media, some recommender systems were developed for Flicker and aimed to perform personalized POI recommendations based on the target user's images. The author topic-based collaborative filtering (ATCF) method is proposed to enable POI recommendations when the target user visits a new place by discovering topics from the images' metadata. Besides, a Visual-enhanced Probabilistic Matrix Factorization model (VPMF) was proposed, which adds visual features of the images into the collaborative filtering model. Some recommender systems have been developed based on Instagram. One of them utilizes an external knowledge base to build relationships between hashtags and perform picture recommendations based on the correlations. Another method is developed to discover topical authorities related to the target user by inferring topical interests from the user's biography, propagating interests over the follower graph, and assigning topics to authorities. Also, CNN-based methods have been proposed to extract the visual features from the images and perform visual content-enhanced POI recommendations.

12.2.5.3 Techniques Used in Social Network Recommender Systems

- Collaborative filtering (CF)

The model-based RS requires a learning phase in advance for finding out the optimal model parameters before making a recommendation. Once the learning phase is finished, the model-based RS can predict the ratings of users very quickly. Among them, the latent factor model (LFM) is very competitive and widely adopted to implement RS, which factorizes the user-item rating matrix into two low-rank matrices: the user feature and item feature matrices. It can alleviate data sparsity using dimensionality reduction techniques and usually produce more accurate recommendations than the memory-based CF approach, while drastically decreasing the memory requirement and computation complexity.

Memory-based and trust-aware collaborative filtering—trust relationships between users have been introduced into RS as an effective approach to overcome the problems of data sparsity and cold start (Chen et al. 2018). The hybrid approach builds an active user's trust network using trust statements between the users to improve the accuracy of similarities between users. One of the core roles of the trusted network is to resolve the neighbor selection between a user's trust statements and its similarity values.

- Content-based filtering

This is based on the user preferences and similarity analysis between items and then recommending to the target users.

Content-based filtering offers support for message filtration. Specifically, users interact with the system via a GUI to set up and manage their FRs/BLs (Thilagavathi and Taarika 2014). Machine learning-based text categorization techniques are used to automatically allot each short text message with a set of categories based on the content. Short text classifier is built for accurate extraction and set of discriminating features in the message. The neural learning model is employed for efficient text classification. In addition, the neural model is enclosed within a hierarchical two-level classification. Short messages are classified as neutral or non-neutral and then are classified based on the appropriateness to each of the considered categories (Thilagavathi and Taarika 2014).

- Hybrid filtering

The idea of this approach is to maximize the potential of collaborative filtering and content-based filtering (Fararni et al. 2021), making the recommender system more robust. Big data systems, deep learning techniques, and social networks can be used to help make the filtering algorithm and recommender system more robust.

- Context-aware recommendation

A context-aware (CA) recommendation system employs user context such as preferences to generate recommendation systems to activate users depending on the specific context of an active user, such as time, location, user mood, user groups, and so on. CA systems use context information as a filter in a recommendations system. Most features of a social network, such as context for the user preference (Chen et al. 2018).

12.2.5.4 Scope of Improvements in Social Network Recommender Systems

- For RS, only recommending popular and highly rated items to the active user often results in better recommendation results. However, the user can also easily obtain such item information from other sources, that is, the actual value of such recommendation is not high. Therefore, a good RS should be able to discover items that are difficult to be found by users spontaneously, but meanwhile which also fit the users' interests (Chen et al. 2018).
- Recommending items to users relying solely on accuracy not only wastes resources but also brings little benefit. If they cannot explain the recommended results well, then they cannot determine whether the recommended items meet the needs of users, resulting in reduced system reliability. If RS can provide some explanation information when generating recommendations, the reliability of the recommended results may greatly be improved. Meanwhile, they will greatly arouse the users' attention (Chen et al. 2018).

12.2.5.5 Possible Ideas for Improvement of the Existing Social Network Recommender Systems

- Update the user profile frequently to keep track of the user's latest preference for social networking and collect feedback on using the social media from the user, then change the recommendation to better satisfy the need of the user.

12.3 Summary

Background information on the recommender system is given in this chapter and there are six applications listed (healthcare, security, tourism, e-commerce, e-learning, and social network). Within each application, there are some basic techniques used in the corresponding recommender system, among which collaborative filtering and content-based filtering are the most popular ones. Figures, charts as well as mathematical equations in this chapter may be useful to get a better understanding of the recommender system. There are also some inevitable problems such as a cold start which need to be tackled. Therefore, possible ideas are provided to avoid the problem as much as possible and to maximize the potential of the recommender system.

Think Tank

1. *What are the areas recommender systems are applicable?*
2. *How recommender systems will be applicable in a new area?*

3. *What are the possible ways to provide better recommendation?*
4. *How applications will be benefitted by using recommender systems?*

References

Allen RB (2019) User models: theory, method practice. Int J Man-Mach Stud 32:511–543
Amato F, Moscato V, Picariello A, Sperlí G (2017) Recommendation in social media networks. In: IEEE third international conference on multimedia big data (BigMM), 213–216
Ansari MH, Moradi M, NikRah O, Kambakhsh KM (2016) CodERS: a hybrid recommender system for an E-learning system. In: 2nd international conference of signal processing and intelligent systems (ICSPIS), 1–5
Bayati M, Harounabadi A, Akbari D (2022) Developing a location-based recommender system using collaborative filtering technique in the tourism industry. Tehnički glas 16(1)
Chen R, Hua Q, Zhang L, Kong X (2018) A survey of collaborative filtering-based recommender systems: from traditional methods to hybrid methods based on social networks. IEEE Access 6
Cordero P, Enciso M, López D et al (2020). A conversational recommender system for diagnosis using fuzzy rules. Expert Syst Appl
Das D, Sahoo L, Datta S (2016) A survey on recommender system. (IJCSIS) Int J Comput Sci Inf Secur 14(5)
Duo L, Su JT (2015) A recommender system based on contextual information of click and purchase data to items for e-commerce. In: Third international conference on cyberspace technology (CCT 2015)
Fararni KA, Nafis F, Aghoutane B et al (2021) Hybrid recommender system for tourism based on big data and AI: a conceptual framework. Big Data Min Analyt 4(1):47–55
Felix G, Falko T, Jochen S et al (2020) A pharmaceutical therapy recommender system enabling shared decision-making. User Model User-Adapted Interact
Fu CJ, Leng ZH (2018) A framework for recommender systems in E-commerce based on distributed storage and data mining. Int Conf E-Bus E-Govern
Haider MH, Al-Azawei A, Al-A'araji N (2019) Developing a healthcare recommender system using an enhanced symptoms-based collaborative filtering technique. J Comput Theor Nanosci 16:920–926
Hidayatullah A, Anugerah MA (2018) A recommender system for E-commerce using multi-objective ranked bandits algorithm. In: International conference on computing, engineering, and design (ICCED), 170–174
Isinkaye FO, Folajimi YO, Ojokuh BA (2015) Recommender systems: principles, methods and evaluation. Egypt Inform J 16:261–273
Kamran M, Javed A (2015) A survey of recommender systems and their application in healthcare. Techn J 20(4). https://prdb.pk/article/a-survey-of-recommender-systems-and-their-application-in-hea-7462
Katarya R, Verma O (2017) Privacy-preserving and secure recommender system enhance with K-NN and social tagging, 52–57. https://doi.org/10.1109/CSCloud.2017.24
Kunaver M, Požrl T (2017) Diversity in recommender systems—a survey. Knowl-Based Syst 123:154–162
Lu J et al (2015) Recommender system application developments: a survey. Decis Support Syst 74:12–32
Lu J, Zhang Q, Zhang GQ (2020) Recommender systems in intelligent information system, 6

Nagaraj P, Deepalakshmi P (2019) A framework for e-healthcare management service using recommender system. Electron Gov 16(1–2):84–100
Nath AS (2018) A pragmatic review on different approaches used in E-learning recommender systems. In: International conference on circuits and systems in digital enterprise technology (ICCSDET), 1–4
Ouaftouh S, Sassi I, Zellou, A, Anter S (2019) Flat and hierarchical user profile clustering in an e-commerce recommender system. In: 1st international conference on smart systems and data science (ICSSD), 1–5
Sahoo AK, Pradhan C, Barik RK, Dubey H (2019) DeepReco: deep learning based health recommender system using collaborative filtering. Computation 7(2):25
Sondess M, Faten K, Marco V et al (2019) LOOKER: a mobile, personalized recommender system in the tourism domain based on social media user-generated content. Pers Ubiquit Comput 23:181–197
Stark B, Knahl C, Aydin M, Karim (2019) A literature review on medicine recommender. Int J Adv Comput Sci Appl (IJACSA) 10(8)
Sweeney L (2020) K-anonymity: a model for protecting privacy. Internat J Uncertain Fuzziness Knowl-Based Syst 10(05):557–570
Thilagavathi N, Taarika R (2014) Content based filtering in online social network using inference algorithm. In: International conference on circuits, power and computing technologies [ICCPCT-2014], 1416–1420
Tran TNT et al (2020) Recommender systems in the healthcare domain: state-of-the-art and research issues. J Intell Inf Syst 57(3):171–201
Weiming H, Baisong L, Hao T (2019) Privacy protection for recommendation system: a survey. J Phys Conf Ser 1325:012087. https://doi.org/10.1088/1742-6596/1325/1/012087
Wong RCW, Li J, Fu AWC et al (2006) (α, k)-anonymity: an enhanced k-anonymity model for privacy preserving data publishing. In: Proceedings of the 12th ACM SIGKDD international conference on knowledge discovery and data mining, 754–759
Yubo J, Hao C, Chengwei H (2010) A collaborative filtering recommendation algorithm based on user trust model. In: First international conference on networking and distributed computing, 213–217
Zhang X, Wang L (2013) The application of web log in collaborative filtering recommendation algorithm. In: Ninth international conference on computational intelligence and security, 763–765
Zhao X (2019) A study on E-commerce recommender system based on big data. In: IEEE 4th international conference on cloud computing and big data analysis (ICCCBDA), 222–226
Zhao X, Ji K (2013) Tourism E-commerce recommender system based on web data mining. In: 8th international conference on computer science & education, 1485–1488
Zheng XY, Luo YL, Sun LZ, Ji Z, Chen FL (2018) A tourism destination recommender system using users' sentiment and temporal dynamics. J Intell Inf Syst 51:557–558

Index

P. Kar et al., *Recommender Systems: Algorithms and their Applications*, Transactions on Computer Systems and Networks, https://doi.org/10.1007/978-981-97-0538-2

MIX
Papier aus verantwortungsvollen Quellen
Paper from responsible sources
FSC® C105338

If you have any concerns about our products,
you can contact us on
ProductSafety@springernature.com

In case Publisher is established outside the EU,
the EU authorized representative is:
Springer Nature Customer Service Center GmbH
Europaplatz 3, 69115 Heidelberg, Germany

Printed by Libri Plureos GmbH
in Hamburg, Germany